The Scientific Context for

EXPLORATION
of the
MOON

Committee on the Scientific Context for Exploration of the Moon
Space Studies Board
Division on Engineering and Physical Sciences

NATIONAL RESEARCH COUNCIL
OF THE NATIONAL ACADEMIES

THE NATIONAL ACADEMIES PRESS
Washington, D.C.
www.nap.edu

THE NATIONAL ACADEMIES PRESS   **500 Fifth Street, N.W.**   **Washington, DC 20001**

NOTICE: The project that is the subject of this report was approved by the Governing Board of the National Research Council, whose members are drawn from the councils of the National Academy of Sciences, the National Academy of Engineering, and the Institute of Medicine. The members of the committee responsible for the report were chosen for their special competences and with regard for appropriate balance.

This study is based on work supported by the Contract NASW-010001 between the National Academy of Sciences and the National Aeronautics and Space Administration. Any opinions, findings, conclusions, or recommendations expressed in this publication are those of the author(s) and do not necessarily reflect the views of the agency that provided support for the project.

International Standard Book Number-13: 978-0-309-10919-2
International Standard Book Number-10: 0-309-10919-1

Cover: Design by Penny E. Margolskee. All images courtesy of the National Aeronautics and Space Administration.

Copies of this report are available free of charge from:

Space Studies Board
National Research Council
500 Fifth Street, N.W.
Washington, DC 20001

Additional copies of this report are available from the National Academies Press, 500 Fifth Street, N.W., Lockbox 285, Washington, DC 20055; (800) 624-6242 or (202) 334-3313 (in the Washington metropolitan area); Internet, http://www.nap.edu.

Printed in the United States of America

THE NATIONAL ACADEMIES
Advisers to the Nation on Science, Engineering, and Medicine

The **National Academy of Sciences** is a private, nonprofit, self-perpetuating society of distinguished scholars engaged in scientific and engineering research, dedicated to the furtherance of science and technology and to their use for the general welfare. Upon the authority of the charter granted to it by the Congress in 1863, the Academy has a mandate that requires it to advise the federal government on scientific and technical matters. Dr. Ralph J. Cicerone is president of the National Academy of Sciences.

The **National Academy of Engineering** was established in 1964, under the charter of the National Academy of Sciences, as a parallel organization of outstanding engineers. It is autonomous in its administration and in the selection of its members, sharing with the National Academy of Sciences the responsibility for advising the federal government. The National Academy of Engineering also sponsors engineering programs aimed at meeting national needs, encourages education and research, and recognizes the superior achievements of engineers. Dr. Charles M. Vest is president of the National Academy of Engineering.

The **Institute of Medicine** was established in 1970 by the National Academy of Sciences to secure the services of eminent members of appropriate professions in the examination of policy matters pertaining to the health of the public. The Institute acts under the responsibility given to the National Academy of Sciences by its congressional charter to be an adviser to the federal government and, upon its own initiative, to identify issues of medical care, research, and education. Dr. Harvey V. Fineberg is president of the Institute of Medicine.

The **National Research Council** was organized by the National Academy of Sciences in 1916 to associate the broad community of science and technology with the Academy's purposes of furthering knowledge and advising the federal government. Functioning in accordance with general policies determined by the Academy, the Council has become the principal operating agency of both the National Academy of Sciences and the National Academy of Engineering in providing services to the government, the public, and the scientific and engineering communities. The Council is administered jointly by both Academies and the Institute of Medicine. Dr. Ralph J. Cicerone and Dr. Charles M. Vest are chair and vice chair, respectively, of the National Research Council.

www.national-academies.org

OTHER REPORTS OF THE SPACE STUDIES BOARD

Earth Science and Applications from Space: National Imperatives for the Next Decade and Beyond (2007)
Exploring Organic Environments in the Solar System (SSB with the Board on Chemical Sciences and Technology, 2007)
A Performance Assessment of NASA's Astrophysics Program (SSB with the Board on Physics and Astronomy, 2007)

An Assessment of Balance in NASA's Science Programs (2006)
Assessment of NASA's Mars Architecture 2007-2016 (2006)
Assessment of Planetary Protection Requirements for Venus Missions: Letter Report (2006)
Distributed Arrays of Small Instruments for Solar-Terrestrial Research: Report of a Workshop (2006)
Issues Affecting the Future of the U.S. Space Science and Engineering Workforce: Interim Report (SSB with the Aeronautics and Space Engineering Board [ASEB], 2006)
Review of NASA's 2006 Draft Science Plan: Letter Report (2006)
The Scientific Context for Exploration of the Moon: Interim Report (2006)
Space Radiation Hazards and the Vision for Space Exploration (2006)

The Astrophysical Context of Life (SSB with the Board on Life Sciences, 2005)
Earth Science and Applications from Space: Urgent Needs and Opportunities to Serve the Nation (2005)
Extending the Effective Lifetimes of Earth Observing Research Missions (2005)
Preventing the Forward Contamination of Mars (2005)
Principal-Investigator-Led Missions in the Space Sciences (2005)
Priorities in Space Science Enabled by Nuclear Power and Propulsion (SSB with ASEB, 2005)
Review of Goals and Plans for NASA's Space and Earth Sciences (2005)
Review of NASA Plans for the International Space Station (2005)
Science in NASA's Vision for Space Exploration (2005)

Assessment of Options for Extending the Life of the Hubble Space Telescope: Final Report (SSB with ASEB, 2004)
Exploration of the Outer Heliosphere and the Local Interstellar Medium: A Workshop Report (2004)
Issues and Opportunities Regarding the U.S. Space Program: A Summary Report of a Workshop on National Space Policy (SSB with ASEB, 2004)
Plasma Physics of the Local Cosmos (2004)
Review of Science Requirements for the Terrestrial Planet Finder: Letter Report (2004)
Understanding the Sun and Solar System Plasmas: Future Directions in Solar and Space Physics (2004)
Utilization of Operational Environmental Satellite Data: Ensuring Readiness for 2010 and Beyond (SSB with ASEB and the Board on Atmospheric Sciences and Climate, 2004)

Limited copies of these reports are available free of charge from:

Space Studies Board
National Research Council
The Keck Center of the National Academies
500 Fifth Street, N.W., Washington, DC 20001
(202) 334-3477/ssb@nas.edu
www.nationalacademies.org/ssb/ssb.html

NOTE: Listed according to year of approval for release, which in some cases precedes the year of publication.

COMMITTEE ON THE SCIENTIFIC CONTEXT FOR EXPLORATION OF THE MOON

GEORGE A. PAULIKAS, The Aerospace Corporation (retired), *Chair*
CARLÉ M. PIETERS, Brown University, *Vice Chair*
WILLIAM B. BANERDT, Jet Propulsion Laboratory
JAMES L. BURCH, Southwest Research Institute
ANDREW CHAIKIN, Science Journalist, Arlington, Vermont
BARBARA A. COHEN, University of New Mexico
MICHAEL DUKE,[1] Colorado School of Mines
ANTHONY W. ENGLAND,[2] University of Michigan
HARALD HIESINGER, Westfälische Wilhelms-Universität, Münster
NOEL W. HINNERS, University of Colorado
AYANNA M. HOWARD, Georgia Institute of Technology
DAVID J. LAWRENCE, Los Alamos National Laboratory
DANIEL F. LESTER, McDonald Observatory
PAUL G. LUCEY, University of Hawaii
S. ALAN STERN,[3] Southwest Research Institute
STEFANIE TOMPKINS, Science Applications International Corporation
FRANCISCO P.J. VALERO, Scripps Institution of Oceanography
JOHN W. VALLEY, University of Wisconsin-Madison
CHARLES D. WALKER, Independent Consultant, Annandale, Virginia
NEVILLE J. WOOLF, University of Arizona

Staff

ROBERT L. RIEMER, Study Director
DAVID H. SMITH, Senior Staff Officer
RODNEY N. HOWARD, Senior Project Assistant
CATHERINE A. GRUBER, Assistant Editor
STEPHANIE BEDNAREK, Research Assistant

---

[1]During committee deliberations, Dr. Duke recused himself from discussion of the finding and recommendation related to the South Pole-Aitken Basin.

[2]Dr. England resigned from the committee on August 11, 2006, because of other commitments.

[3]Dr. Stern resigned from the committee on September 24, 2006, to join the NASA Advisory Committee Science Subcommittee (and on April 2, 2007, became Associate Administrator for NASA's Science Mission Directorate).

Preface

As an initial part of the nation's newly established Vision for Space Exploration,[1] the National Aeronautics and Space Administration (NASA) is planning missions to the Moon through the Exploration Systems Mission Directorate (ESMD). The first of these NASA missions, the Lunar Reconnaissance Orbiter, is already in implementation and scheduled for a 2008 launch.

Looking beyond the several lunar robotic missions to be flown by 2008 (by the international community), science goals need to be articulated for early decisions about system design and operations planning for later robotic and human activities on the Moon. For a longer-range human presence on the Moon, the scope of science is potentially broader, including extensive field studies and sampling, plus the emplacement or assembly and the maintenance and operation of major equipment on the lunar surface. After a substantial hiatus in lunar science and exploration activities, the next two decades will be marked by a major resurgence in lunar missions and high potential for scientific return. In order to realize this benefit from the initial series of missions, NASA needs a comprehensive, well-validated, and prioritized set of scientific research objectives for a program of exploration of the Moon. The purpose of this report is to provide scientific input to NASA's planning process.

This study was initiated at the request of Mary Cleave, NASA's associate administrator for science, in a letter dated March 13, 2006, to Lennard Fisk, chair of the Space Studies Board (SSB), asking the National Research Council (NRC) to provide guidance on the scientific challenges and opportunities enabled by a sustained program of robotic and human exploration of the Moon during the period 2008-2023 and beyond.

In response to this request and to meet the ambitious schedule requested by NASA, the NRC established the Committee on the Scientific Context for Exploration of the Moon (biographies of the committee members appear in Appendix F) in May 2006. The committee met at the Keck Center of the National Academies in Washington, D.C., on June 20-22, 2006, and at the Beckman Center, Irvine, California, on August 2-4, 2006. An interim report, requested by NASA, was delivered to NASA in mid-September 2006.[2] Subsequently, the committee met in Santa Fe, New Mexico, on October 25-27, 2006, and in Boulder, Colorado, on February 13-15, 2007. The agendas of these meetings are presented in Appendix C. In addition, committee members consulted related reports issued by the National Research Council (listed in the Bibliography).

---

[1]National Aeronautics and Space Administration (NASA), The Vision for Space Exploration, NP-2004-01-334-HQ, NASA, Washington, D.C., 2004.

[2]National Research Council, The Scientific Context for Exploration of the Moon: Interim Report, The National Academies Press, Washington, D.C., 2006.

The committee held several teleconference calls, communicated extensively via e-mail, and solicited input from colleagues with expertise relevant to the study of the Moon and/or the development and operation of spaceflight instrumentation and robotic spacecraft. The committee, encouraged by NASA to reach out to the broad scientific community, also presented the results of the interim report and its plans for this final report at several venues in the United States and abroad. A summary of the locations and audiences for the outreach presentations is given in Appendix E.

The work of the committee was made easier thanks to the important help, advice, and comments provided by numerous individuals from a variety of public and private organizations. The committee heard presentations from the following NASA staff, university researchers, and other experts: Rob Ambrose, NASA JSC; Joseph Borovsky, Los Alamos National Laboratory; Jack O. Burns, University of Colorado, Boulder; Gordon Chin, NASA GSFC; Robert Fogel, NASA SMD; James Head III, Brown University; Paul Hertz, NASA Science Mission Directorate; Butler P. Hine III, NASA Ames Research Center; Brad Jolliff, Washington University at St. Louis; David Lavery, NASA SMD; Mario Livio, Space Telescope Science Institute; Gary Lofgren, NASA Johnson Space Center; Clive R. Neal, University of Notre Dame; Charles Shearer, University of New Mexico; Norman Sleep, Stanford University; Paul Spudis, Johns Hopkins University; Timothy Stubbs, University of Maryland; G. Jeffrey Taylor, University of Hawaii; S. Ross Taylor, Australian National University; Richard R. Vondrak, NASA GSFC; Michael Wargo, NASA Exploration Systems Mission Directorate; Simon P. Worden, NASA Ames Research Center; and committee members James Burch, Southwest Research Institute; Noel Hinners, University of Colorado; Ayanna Howard, Georgia Institute of Technology; Daniel Lester, McDonald Observatory; Francisco Valero, Scripps Institute of Oceanography; John W. Valley, University of Wisconsin-Madison; and Neville J. Woolf, University of Arizona.

In addition to the above speakers, the following individuals and groups provided useful input to the committee: David Beaty, Paul Schenker, and Edward W. Tunstel, Jet Propulsion Laboratory; Donald Bogard, Friedrich Horz, John Jones, and Sarah Noble, NASA Johnson Space Center; Jack O. Burns, University of Colorado, Boulder; Ian A. Crawford, Birkbeck College, United Kingdom; Lisa Gaddis, U.S. Geological Survey, Flagstaff; Rick Halbach, Lockheed Martin Corporation; William Hartmann, Planetary Sciences Institute; Lon Hood, University of Arizona; Boris Ivanov, Russian Academy of Sciences; Jonathan Levine, University of Chicago; the Moon-Mars Science Linkages Science Steering Group of the Mars Exploration Program Assessment Group; Noah Petro, Brown University; Harrison H. Schmitt, NASA Advisory Council; John Stevens, Lockheed Martin Corporation; Robert Strom, University of Arizona; Timothy Swindle, University of Arizona; and Lawrence Taylor, University of Tennessee. We thank Bruce Jakosky, Ariel Anbar, Jeffrey Taylor, and Paul Lucey for their paper on astrobiology and lunar exploration; Clive R. Neal, Lon Hood, Shaopeng Huang, and Yosio Nakamura for their white paper "Scientific Rationale for Deployment of a Long Lived Geophysical Network on the Moon"; Timothy Stubbs, Richard Vondrak, and William Farrel for "A Dynamic Fountain Model for Lunar Dust"; and contributors, too numerous to list, in a Lunar Exploration Analysis Group (LEAG) report on lunar science.

The committee also thanks SSB research assistant Stephanie Bednarek for her valuable assistance in assembling the draft of the interim report and assisting at the committee's meetings.

This report has been reviewed in draft form by individuals chosen for their diverse perspectives and technical expertise, in accordance with procedures approved by the NRC's Report Review Committee. The purpose of this independent review is to provide candid and critical comments that will assist the institution in making its published report as sound as possible and to ensure that the report meets institutional standards for objectivity, evidence, and responsiveness to the study charge. The review comments and draft manuscript remain confidential to protect the integrity of the deliberative process. We wish to thank the following individuals for their review of this report:

Ariel Anbar, Arizona State University,
Rodney A. Brooks, Massachusetts Institute of Technology,
I.A. Crawford, University of London,
Tamara E. Jernigan, Lawrence Livermore National Laboratory,
Ian Pryke, Center for Aerospace Policy Research, George Mason University,
Richard J. Robbins, The Robbins Group LLC,
Irwin Shapiro, Harvard-Smithsonian Center for Astrophysics,

Harlan E. Spence, Boston University,
Lawrence A. Taylor, University of Tennessee, and
Mark Wieczorek, Institut de Physique du Globe de Paris.

Although the reviewers listed above have provided many constructive comments and suggestions, they were not asked to endorse the conclusions or recommendations, nor did they see the final draft of the report before its release. The review of this report was overseen by Bernard F. Burke, Massachusetts Institute of Technology, and William G. Agnew, General Motors Corporation (retired). Appointed by the NRC, they were responsible for making certain that an independent examination of this report was carried out in accordance with institutional procedures and that all review comments were carefully considered. Responsibility for the final content of this report rests entirely with the authoring committee and the institution.

George A. Paulikas, *Chair*, and
Carlé M. Pieters, *Vice Chair*
Committee on the Scientific Context
for Exploration of the Moon

Contents

Executive Summary

We know more about many aspects of the Moon than about any world beyond our own, and yet we have barely begun to solve its countless mysteries. The Moon is, above all, a witness to 4.5 billion years (Ga) of solar system history, and it has recorded that history more completely and more clearly than has any other planetary body. Nowhere else can we see back with such clarity to the time when Earth and the other terrestrial planets were formed and life emerged on Earth.

Planetary scientists have long understood the Moon's unique place in the evolution of rocky worlds. Many of the processes that have modified the terrestrial planets have been absent on the Moon. The lunar interior retains a record of the initial stages of planetary evolution. Its crust has never been altered by plate tectonics, which continually recycle Earth's crust; or by planetwide volcanism, which resurfaced Venus only half a billion years ago; or by the action of wind and water, which have transformed the surfaces of both Earth and Mars. The Moon today presents a record of geologic processes of early planetary evolution in the purest form.

Lunar science provides a window into the early history of the Earth-Moon system, can shed light on the evolution of other terrestrial planets such as Mars and Venus, and can reveal the record of impacts within the inner solar system. By dint of its proximity to Earth, the Moon is accessible to a degree that other planetary bodies are not.

For these reasons, the Moon is priceless to planetary scientists. It remains a cornerstone for deciphering the histories of those more complex worlds. But because of the limitations of current data, researchers cannot be sure that they have read these histories correctly. Now, thanks to the legacy of the Apollo program and subsequent missions, such as Clementine and Lunar Prospector, and looking forward to the newly established Vision for Space Exploration (VSE),[1] scientists are able to pose sophisticated questions that are more relevant and focused than those that could be asked over three decades ago. Only by returning to the Moon to carry out new scientific explorations can we hope to narrow the gaps in understanding and learn the secrets that the Moon alone has kept for eons.

The Moon is not only of intrinsic interest as a cornerstone of the Earth-Moon system science, but it also provides a unique location for research in several other fields of science. The Moon's surface is in direct contact with the interplanetary medium, and the interaction of the Moon with the solar wind plasma flowing from the Sun forms a unique plasma physics laboratory. Astronomical and astrophysical observations as well as observations of Earth, its atmosphere, ionosphere, and magnetosphere may be made from the stable platform of the Moon. The absence of a significant ionosphere on the Moon should enable low-frequency radio astronomy to be carried out, particularly from the farside of the Moon where radio interference from terrestrial sources should be absent.

---

[1]National Aeronautics and Space Administration (NASA), *The Vision for Space Exploration*, NP-2004-01-334-HQ, NASA, Washington, D.C., 2004.

NASA asked the National Research Council (NRC) to provide guidance on the scientific challenges and opportunities enabled by a sustained program of robotic and human exploration of the Moon during the period 2008-2023 and beyond as the VSE evolves. This report was prepared by the Committee on the Scientific Context for Exploration of the Moon (brief biographies of the committee are presented in Appendix F).

The framework of the VSE was changing while this report was being prepared. However, the committee believes that its scientific rationale for lunar science and its goals and recommendations are independent of any particular programmatic implementation.

It is the unanimous consensus of the committee that the Moon offers profound scientific value. The infrastructure provided by sustained human presence can enable remarkable science opportunities if those opportunities are evaluated and designed into the effort from the outset. While the expense of human exploration cannot likely be justified on the basis of science alone, the committee emphasizes that careful attention to the science opportunity is very much in the interest of a stable and sustainable lunar program. In the opinion of the committee, a vigorous near-term robotic exploration program providing global access is central to the next phase of scientific exploration of the Moon and is necessary both to prepare for the efficient utilization of human presence and to maintain scientific momentum as this major national program moves forward.

PRIORITIES, FINDINGS, AND RECOMMENDATIONS

According to the committee's statement of task (see Appendix A):

> The current study is intended to meet the near-term needs for science guidance for the lunar component of the VSE. . . . [T]he *primary goals* of the study are to:
> 1. Identify a common set of prioritized basic science goals that could be addressed in the near-term via the LPRP[2] program of orbital and landed robotic lunar missions (2008-2018) and in the early phase of human lunar exploration (nominally beginning in 2018); and
> 2. To the extent possible, suggest whether individual goals are most amenable to orbital measurements, in situ analysis or instrumentation, field observation or terrestrial analysis via documented sample return.

Also outlined in the statement of task are the overall science scope for this study and several secondary tasks.

Overarching Themes

The committee identified four overarching themes of lunar science: early Earth-Moon system, terrestrial planet differentiation and evolution, solar system impact record, and lunar environment. The committee then constructed eight science concepts that address broad areas of scientific research. Each is multicomponent and is linked to different aspects of the overarching themes of lunar science.

The committee approached the challenge of prioritization by developing a hierarchy of priority categories. It used the prioritization criteria adopted by the decadal survey *New Frontiers in the Solar System: An Integrated Exploration Strategy*[3] as a guideline: the criteria are scientific merit, opportunity, and technological readiness.

The committee thus structured the prioritization of goals called for in the statement of task along three lines: (1) prioritization of science concepts, (2) prioritization of science goals, and (3) specific integrated high-priority recommendations. Although the rationales for these three are linked throughout the discussion of this report, the implementation requirements are different. As requested in the statement of task, the priorities and recommendations presented in this report relate to the near-term implementation of the VSE, which includes the robotic precursors and initial human excursions on the Moon. Planning for and implementing longer-term scientific activities on the Moon are beyond the scope of this study.

---

[2]The Lunar Precursor and Robotic Program (LPRP) was how robotic missions were identified in the NASA letter that requested this study. The LPRP terminology is no longer in use.

[3]National Research Council, *New Frontiers in the Solar System: An Integrated Exploration Strategy*, The National Academies Press, Washington, D.C., 2003.

Prioritized Science Concepts

The committee evaluated only the scientific merit of each science concept in order to rank the concepts. It should be noted that *all* concepts discussed are viewed to be scientifically important. The science concepts are prioritized below and discussed in more detail in Chapter 3.

1. The bombardment history of the inner solar system is uniquely revealed on the Moon.
2. The structure and composition of the lunar interior provide fundamental information on the evolution of a differentiated planetary body.
3. Key planetary processes are manifested in the diversity of lunar crustal rocks.
4. The lunar poles are special environments that may bear witness to the volatile flux over the latter part of solar system history.
5. Lunar volcanism provides a window into the thermal and compositional evolution of the Moon.
6. The Moon is an accessible laboratory for studying the impact process on planetary scales.
7. The Moon is a natural laboratory for regolith processes and weathering on anhydrous airless bodies.
8. Processes involved with the atmosphere and dust environment of the Moon are accessible for scientific study while the environment remains in a pristine state.

Prioritization of Science Goals

Within the 8 science concepts above, the committee identified 35 specific science goals that can be addressed, at least in part, during the early phases of the VSE. For these science goals, the committee evaluated science merit as well as the degree to which they can be achieved within current or near-term technical readiness and practical accessibility. Within their respective science concepts, the science goals are listed in the order of their overall priority ranking (a through e) in Table 3.1 in Chapter 3.

All 35 specific science goals were also evaluated and ranked as a group, separately from the science concepts with which they are associated. The highest-ranking lunar science goals are listed in Table 5.1 in Chapter 5 in priority order. For this group of goals the committee identifies possible means of implementation to achieve each goal.

FINDINGS AND RECOMMENDATIONS

Principal Finding: Lunar activities apply to broad scientific and exploration concerns.

Lunar science as described in this report has much broader implications than simply studying the Moon. For example, a better determination of the lunar impact flux during early solar system history would have profound implications for comprehending the evolution of the solar system, early Earth, and the origin and early evolution of life. A better understanding of the lunar interior would bear on models of planetary formation in general and on the origin of the Earth-Moon system in particular. And exploring the possibly ice-rich lunar poles could reveal important information about the history and distribution of solar system volatiles. Furthermore, although some of the committee's objectives are focused on lunar-specific questions, one of the basic principles of comparative planetology is that each world studied enables researchers to better understand other worlds, including our own. Improving our understanding of such processes as cratering and volcanism on the Moon will provide valuable points of comparison for these processes on the other terrestrial planets.

Finding 1: Enabling activities are critical in the near term.

A deluge of spectacular new data about the Moon will come from four sophisticated orbital missions to be launched between 2007 and 2008: SELENE (Japan), Chang'e (China), Chandrayaan-1 (India), and the Lunar Reconnaissance Orbiter (United States). Scientific results from these missions, integrated with new analyses of existing data and samples, will provide the enabling framework for implementing the VSE's lunar activities. However, NASA and the scientific community are currently underequipped to harvest these data and produce meaningful information. For example, the lunar science community assembled at the height of the Apollo program of the late 1960s and early 1970s has since been depleted in terms of its numbers and expertise base.

> **Recommendation 1a:** NASA should make a strategic commitment to stimulate lunar research and engage the broad scientific community[4] by establishing two enabling programs, one for fundamental lunar research and one for lunar data analysis. Information from these two recommended efforts—a Lunar Fundamental Research Program and a Lunar Data Analysis Program—would speed and revolutionize understanding of the Moon as the Vision for Space Exploration proceeds.

> **Recommendation 1b:** The suite of experiments being carried by orbital missions in development will provide essential data for science and for human exploration. NASA should be prepared to recover data lost due to failure of missions or instruments by reflying those missions or instruments where those data are deemed essential for scientific progress.

Finding 2: Strong ties with international programs are essential.

The current level of planned and proposed activity indicates that almost every space-faring nation is interested in establishing a foothold on the Moon. Although these international thrusts are tightly coupled to technology development and exploration interests, science will be a primary immediate beneficiary. NASA has the opportunity to provide leadership in this activity, an endeavor that will remain highly international in scope.

> **Recommendation 2:** NASA should explicitly plan and carry out activities with the international community for scientific exploration of the Moon in a coordinated and cooperative manner. The committee endorses the concept of international activities as exemplified by the recent "Lunar Beijing Declaration" of the 8th ILEWG (International Lunar Exploration Working Group) International Conference on Exploration and Utilization of the Moon (see Appendix D).

Finding 3: Exploration of the South Pole-Aitken Basin remains a priority.

The answer to several high-priority science questions identified can be found within the South Pole-Aitken Basin, the oldest and deepest observed impact structure on the Moon and the largest in the solar system. Within it lie samples of the lower crust and possibly the lunar mantle, along with answers to questions on crater and basin formation, lateral and vertical compositional diversity, lunar chronology, and the timing of major impacts in the early solar system.

Missions to South Pole-Aitken Basin, beginning with robotic sample returns and continuing with robotic and human exploration, have the potential to be a cornerstone for lunar and solar system research. (A South Pole-Aitken Basin sample-return mission was listed as a high priority in the 2003 NRC decadal survey report *New Frontiers in the Solar System: An Integrated Exploration Strategy*.[5])

> **Recommendation 3:** NASA should develop plans and options to accomplish the scientific goals set out in the high-priority recommendation in the National Research Council's *New Frontiers in the Solar System: An Integrated Exploration Strategy* (2003) through single or multiple missions that increase understanding of the South Pole-Aitken Basin and by extension all of the terrestrial planets in our solar system (including the timing and character of the late heavy bombardment).

Finding 4: Diversity of lunar samples is required for major advances.

Laboratory analyses of returned samples provide a unique perspective based on scale, precision, and flexibility of analysis and have permanence and ready accessibility. The lunar samples returned during the Apollo and

---

[4]See also National Research Council, *Building a Better NASA Workforce: Meeting the Workforce Needs for the National Vision for Space Exploration*, The National Academies Press, Washington, D.C., 2007.

[5]National Research Council, *New Frontiers in the Solar System: An Integrated Exploration Strategy*, The National Academies Press, Washington, D.C., 2003.

Luna missions dramatically changed understanding of the character and evolution of the solar system. Scientists now understand, however, that these samples are not representative of the larger Moon and do not provide sufficient detail and breadth to address the fundamental science concepts outlined in Table 3.1 in this report.

Recommendation 4: Landing sites should be selected that can fill in the gaps in diversity of lunar samples. Mission plans for each human landing should include the collection and return of at least 100 kg of rocks from diverse locations within the landing region. For all missions, robotic and human, to improve the probability of finding new, ejecta-derived diversity among smaller rock fragments, every landed mission that will return to Earth should retrieve at least 1 kg of rock fragments 2 to 6 mm in diameter separated from bulk soil. Each mission should also return 100 to 200 grams of unfractionated regolith.

Finding 5: The Moon may provide a unique location for observation and study of Earth, near-Earth space, and the universe.

The Moon is a platform that can potentially be used to make observations of Earth (Earth science) and to collect data for heliophysics, astrophysics, and astrobiology. Locations on the Moon provide both advantages and disadvantages. There are substantial uncertainties in the benefits and the costs of using the Moon as an observation platform as compared with alternate locations in space. The present committee did not have the required span and depth of expertise to perform a thorough evaluation of the many issues that need examination. A thorough study is required.

Recommendation 5: The committee recommends that NASA consult scientific experts to evaluate the suitability of the Moon as an observational site for studies of Earth, heliophysics, astronomy, astrophysics, and astrobiology. Such a study should refer to prior NRC decadal surveys and their established priorities.

RELATED ISSUES

The committee identified several related issues pertaining to optimal implementation of science in the VSE. This effort was driven by the stark realization that more than 30 years have passed since Apollo and that the nature of the VSE itself warrants a major reconsideration of the basic approach to conducting lunar science. In those more than 30 years, robotic capability has increased dramatically, analytical instrumentation has advanced remarkably, and the very understanding of how to explore has evolved as scientists have learned about planetary formation and evolution. The VSE offers new opportunity: there is no longer the limitation of short-duration lunar stays of 2 or 3 days and "emplacement science"; scientists on the Moon can operate as scientists, doing analytical work and deciphering sample/source relationships; site revisit with follow-up science is possible (e.g., an outpost); robotic-capable equipment can be used between missions; geophysical equipment can be used in survey modes; time-consuming deep drilling is possible; high-grade lunar samples can be selected for return to Earth. Nurturing a new approach to lunar exploration must be fostered if the potential of the VSE is to be reaped.

Finding 1R: The successful integration of science into programs of human exploration has historically been a challenge. It remains so for the VSE. Prior Space Studies Board reports by the Committee on Human Exploration (CHEX) examined how the different management approaches led to different degrees of success. CHEX developed principles for optimizing the integration of science into human exploration and recommended implementation of these principles in future programs.[6] This committee adopts in Recommendation 1R the CHEX findings in a form appropriate for the early phase of VSE.

Recommendation 1R: NASA should increase the potential to successfully accomplish science in the VSE by (1) developing an integrated human/robotic science strategy,[7] (2) clearly stating where science fits in the

---

[6]See p. 128 of the third report in a series by the Committee on Human Exploration: National Research Council, *Science Management in the Human Exploration of Space*, National Academy Press, Washington, D.C., 1997.

[7]This CHEX Recommendation 1 refers to the development of science goals, strategy, priorities, and process methodology; CHEX Recommendation 3 (and this committee's Recommendation 1R) refers strictly to the implementation of science in a program of human exploration.

Exploration Systems Mission Directorate's (ESMD's) goals and priorities, and (3) establishing a science office embedded in the ESMD to plan and implement science in the VSE. Following the Apollo model, such an office should report jointly to the Science Mission Directorate and the ESMD, with the science office controlling the proven end-to-end science process.

Finding 2R: Great strides and major advances in robotics, space and information technology, and exploration techniques have been made since Apollo. These changes are accompanied by a greatly evolved understanding of and approach to planetary science and improvements in use of remote sensing and field and laboratory sample analyses. Critical to achieving high science return in Apollo was the selection of the lunar landing sites and the involvement of the science community in that process. Similarly, the scientific community's involvement in detailed mission planning and implementation resulted in efficient and productive surface traverses and instrument deployments.

> **Recommendation 2R:** The development of a comprehensive process for lunar landing site selection that addresses the science goals of Table 5.1 in this report should be started by a science definition team. The choice of specific sites should be permitted to evolve as the understanding of lunar science progresses through the refinement of science goals and the analysis of existing and newly acquired data. Final selection should be done with the full input of the science community in order to optimize the science return while meeting engineering and safety constraints. Similarly, science mission planning should proceed with the broad involvement of the science and engineering communities. The science should be designed and implemented as an integrated human/robotic program employing the best each has to offer. Extensive crew training and mission simulation should be initiated early to help devise optimum exploration strategies.

Finding 3R: The opportunity provided by the VSE to accomplish science, lunar and otherwise, is highly dependent for success on modernizing the technology and instrumentation available. The virtual lack of a lunar science program and no human exploration over the past 30 years have resulted in a severe lack of qualified instrumentation suitable for the lunar environment. Without such instrumentation, the full and promising potential of the VSE will not be realized.

> **Recommendation 3R:** NASA, with the intimate involvement of the science community, should immediately initiate a program to develop and upgrade technology and instrumentation that will enable the full potential of the VSE. Such a program must identify the full set of requirements as related to achieving priority science objectives and prioritize these requirements in the context of programmatic constraints. In addition, NASA should capitalize on its technology development investments by providing a clear path into flight development.

Finding 4R: The NASA curatorial facilities and staff have provided an exemplary capability since the Apollo program to take advantage of the scientific information inherent in extraterrestrial samples. The VSE has the potential to add significant demands on the curatorial facilities. The existing facilities and techniques are not sufficient to accommodate that demand and the new requirements that will ensue. Similarly, there is a need for new approaches to the acquisition of samples on lunar missions.

> **Recommendation 4R:** NASA should conduct a thorough review of all aspects of sample curation, taking into account the differences between a lunar outpost-based program and the sortie approach taken by the Apollo missions. This review should start with a consideration of documentation, collection, and preservation procedures on the Moon and continue to a consideration of the facilities requirements for maintaining and analyzing the samples on Earth. NASA should enlist a broad group of scientists familiar with curatorial capabilities and the needs of lunar science, such as the Curation and Analysis Planning Team for Extraterrestrial Materials (CAPTEM), to assist it with the review.

1

Introduction

WHY LUNAR SCIENCE?

We know more about many aspects of the Moon than we know about any world beyond our own, and yet we have barely begun to solve its countless mysteries. In the decades since the last Apollo landing on the Moon in 1972, there has been a widespread misperception that the Moon has already told us all the important things that it has to tell, that scientifically it is a "been there, done that" world. Nothing could be farther from the truth.

The Moon is, above all, a witness to 4.5 billion years of solar system history, and it has recorded that history more completely and more clearly than has any other planetary body. Nowhere else can we see back with such clarity to the time when Earth and the other terrestrial planets—Mercury, Venus, an Mars—were formed and life emerged on Earth.

Planetary scientists have long understood the Moon's unique significance as the starting point in the continuum of the evolution of rocky worlds. Many of the processes that have modified the terrestrial planets have been absent on the Moon. The lunar interior retains a record of the initial stages of planetary evolution. Its crust has never been altered by plate tectonics, which continually recycle Earth's crust; or by planetwide volcanism, which resurfaced Venus only half a billion years ago; or by the action of wind and water, which have transformed the surfaces of both Earth and Mars. The Moon today presents a record of geologic processes of early planetary evolution in the purest form. Its airless surface also provides a continuous record of solar-terrestrial processes.

For these reasons, the Moon is priceless to planetary scientists: It remains a cornerstone for deciphering the histories of those more complex worlds. But because of the limitations of current samples and data derived from them, researchers cannot be sure that they have read these histories correctly. Now, thanks to the legacy of the Apollo program, it is possible to pose sophisticated questions that are more relevant and focused than those that could be asked more than three decades ago. Only by returning to the Moon to carry out new scientific explorations can we hope to close the gaps in understanding and learn the secrets that the Moon alone has kept for eons.

OVERARCHING THEMES

Through NASA's Vision for Space Exploration (VSE)[1] the nation is embarking on a challenging and inspirational journey to the Moon, Mars, and beyond. This report focuses on the scientific context for exploration of

---

[1]National Aeronautics and Space Administration (NASA), *The Vision for Space Exploration*, NP-2004-01-334-HQ, NASA, Washington, D.C., 2004.

the Moon, especially in the early phases of implementation of the VSE. The exploration of the Moon is a rich and fruitful endeavor with many facets. The scientific context for the lunar science discussed throughout this report encompasses the four following overarching themes (see Figure 1.1), which are fundamentally important to solar system science, including the history of Earth:

- *Early Earth-Moon System:* The compositional and thermal histories of both the Moon and Earth were closely linked 4.5 billion years ago, after which each evolved separately. A prevailing scientific hypothesis is that the Moon was formed from debris of a collision of a Mars-sized body with the early Earth. How, when, and why did the two parts of the Earth-Moon system take different paths and how have they influenced one another?
- *Terrestrial Planet Differentiation and Evolution:* The Moon is a small planetary body that differentiated into crust, mantle, and core within a few hundred million years after formation. A magma ocean hypothesis describes this early process in terms of fractional differentiation of an initial globe-circling ocean of magma. What are the complexities of this fundamental process, and how can the lunar model be used to understand other rocky planets?
- *Solar System Impact Record:* Since its birth 4.5 billion years ago, the Moon has experienced the full force of early and late bombardment of solar system debris. Regarding early bombardment: A terminal cataclysm hypothesis holds that a burst of large impacts occurred on the Moon (and the inner solar system) about 4.0 billion years ago, which, if confirmed, provides important constraints on the evolution of terrestrial planets and the origin and evolution of life on Earth. Regarding late bombardment: After formation of the planets, the frequency of impacts gradually decreased, perhaps punctuated by occasional periods of increased impacts. The early impact record on Earth has been largely destroyed by erosion and plate tectonics, but it is well preserved on the Moon. What is the history of impact events throughout the past 4.0 billion years that is recorded on the Moon?
- *Lunar Environment:* The surface of the Moon is accessible and special. The lunar atmosphere, though tenuous, is the nearest example of a surface boundary exosphere, the most common type of satellite atmosphere

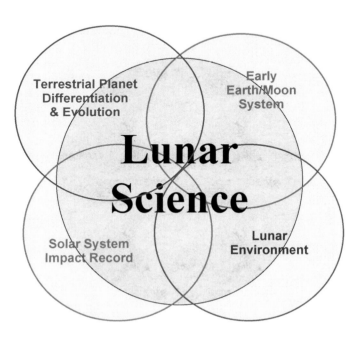

FIGURE 1.1 Lunar science encompasses four overarching themes of solar system exploration.

in the solar system. In a near-vacuum environment with 1/6 the gravity of Earth, the Moon's regolith accumulates products produced over billions of years by exposure to solar and galactic radiation and space plasmas. These products have scientific value and may well have practical value. The frigid lunar poles contain volatiles, perhaps abundant, that may provide information to characterize the sources of volatiles in the early solar system.

All of the overachieving themes are involved in understanding the character and history of the environment at 1 astronomical unit shared by Earth and the Moon. Particularly relevant are implications for the evolution of Earth and the conditions that constrained the formation and evolution of life on Earth.

STRUCTURE OF THIS REPORT

The overarching themes presented above permeate the subjects of Chapters 2 through 5 of this report. An overview of the current understanding of the Earth-Moon system is provided in Chapter 2. The achievements of Apollo era exploration led to several hypotheses about the relation between Earth and the Moon. However, with small amounts of additional data and more sophisticated analytical and computational tools, which have become available in the past few decades, several paradigms have been (or are in the process of being) revised.

The central *science concepts* addressed by lunar exploration are discussed in Chapter 3. Several specific *science goals* that can be addressed in the early phases of the VSE are identified for each concept.

Implementation options and opportunities for addressing the science concepts are summarized in Chapter 4, as are plans of other nations for extensive lunar robotic exploration.

Prioritization criteria for science that can be achieved in the early phases of the VSE are discussed in Chapter 5. While science concepts are prioritized on scientific merit, additional criteria—namely, the availability of opportunities for research and the technological readiness—are used to prioritize individual science goals. Findings and recommendations that envelop and support the prioritization of individual goals are also presented in that chapter.

As the VSE proceeds, the Moon may also provide a unique location for research in several other fields of science, serving as a stable platform for astronomical and astrophysical observations as well as observations of Earth, its atmosphere, ionosphere, and magnetosphere. In addition, there will be opportunities for expanded activities from lunar orbit and at other locations as a result of new launch vehicles. Chapter 6 describes opportunities for research in astronomy and astrophysics and for observations of Earth and its magnetosphere that can take advantage of our return to the Moon.

Since the VSE provides the focus for NASA's activities over the next several decades, there are several additional but related concepts and goals that need attention so as to maximize the efficiency of human and robotic scientific interaction and to optimize the scientific returns from all aspects of lunar research. Issues such as program management and coordination, planning, operations, technology, and the development of facilities all affect the health of the overall VSE undertaking as well as science. Several findings and recommendations related to these issues are offered in Chapter 7. Concluding remarks and the principal finding of the report are presented in Chapter 8.

2

Current Understanding of Early Earth and the Moon

More than 30 years of steady research and explosive technical advances have occurred since the Apollo program dramatically expanded understanding of the origin and evolution of the Moon and its broad significance to planetary science. A set of overarching hypotheses now prevail in lunar science that have broad applicability and implications for planetary science. The maturity of current understanding of the Earth-Moon system has enabled the formulation of tests that, with well-chosen new data, could strengthen, or overturn, some or all of the ruling hypotheses to provide better insight into the character of our solar system.

THE MOON SINCE APOLLO: MAJOR HYPOTHESES AND ENABLING FACTORS

Three hypotheses provide a context for understanding the origin and evolution of the Moon:

- *The giant impact hypothesis* explains the origin of the Moon as being assembled from debris after the impact of a Mars-sized object with the early Earth. (An artist's conception of a giant impact is shown in Figure 2.1.)
- *The lunar magma ocean hypothesis* governs understanding of the formation of lunar rocks following lunar formation, and suggests that the outer portions of the Moon (several hundred kilometers in depth) were entirely molten. Differentiation of the vast magma body, a magma ocean, resulted in the formation of the earliest crust and mantle and produced the rocks observed today.
- *The terminal cataclysm* (sometimes called the Late Heavy Bombardment) *hypothesis* concerns the timing of the impact flux in the 600 million years (Ma) after lunar formation. It proposes that the largest craters observed on the Moon, vast multi-ringed impact basins (e.g., see Figure 2.2), were formed in a brief pulse of impacts of large objects near 4 billion years (Ga) ago, well after impact-causing debris left over from solar system formation had died away (see Figure 2.3). The reality or not of an inner solar system cataclysm is important in understanding conditions on Earth at the time that life was first emerging. An alternate hypothesis is that the rate of impacts to the Moon and Earth declined with time and no cataclysm occurred (also in Figure 2.3).

In addition to these well-formulated hypotheses, there is an emerging recognition that the lunar plasma environment, tenuous atmosphere, regolith, and polar regions in permanent shade constitute a single system in dynamic flux that links the interior of the Moon with the space environment and the volatile history of the solar system.

Since Apollo, several key factors have enabled the formulation of these unifying hypotheses: (1) time for the scientific integration of the large body of disparate data, (2) improved and expanded analytical capability that

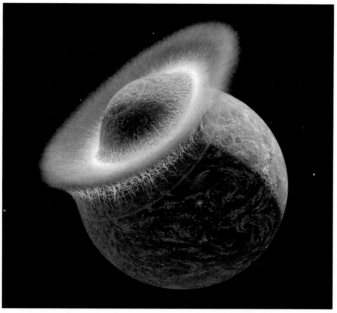

FIGURE 2.1 An artist's conception of a giant impact. The Moon is hypothesized to have been formed at ~4.5 Ga when a Mars-size body called Thea struck Earth. Courtesy of Don Davis.

FIGURE 2.2 Image of the western limb of the Moon obtained by the Galileo spacecraft during an Earth flyby. The Orientale multi-ringed basin is in the center. On the right is the nearside and extensive Mare Procellarum basalts. On the left is the farside, with the South Pole-Aitken Basin dominating the lower left half of the image. Orientale is the youngest basin on the Moon, and South Pole-Aitken Basin is the oldest. SOURCE: NASA Image PIA00077, obtained by the Galileo spacecraft.

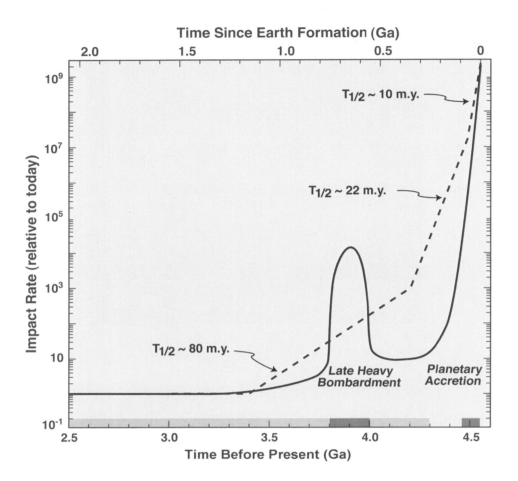

FIGURE 2.3 Competing models of meteorite-impact rate for the first 2 billion years (Ga) of Earth and Moon history. Note that Earth is believed to have formed about 4.55 Ga before present. Two hypotheses are shown: exponential decay of impact rate (dashes; data from W.K. Hartmann, G. Ryder, L. Dones, and D. Grinspoon, The time-dependent intense bombardment of the primordial Earth/Moon system, pp. 493-512 in *Origin of the Earth and Moon* (R.M. Canup and K. Righter, eds.), University of Arizona Press, Tucson, 2000), and cool early Earth–late heavy bombardment (solid curve). The approximate half-life is given in million years (m.y.) for periods of exponential decline. The cool early Earth hypothesis (solid curve) suggests that impact rates had dropped precipitously by 4.4 to 4.3 Ga, consistent with clement conditions that were hospitable for life. SOURCE: Courtesy of John Valley; adapted from J.W. Valley, W.H. Peck, E.M. King, and S.A. Wilde, A cool early Earth, *Geology* 30:351-354, 2002.

exploited the well-documented and curated collection of lunar samples and meteorites, (3) the increase in computational capability, (4) the recognition of meteorites from the Moon, and (5) remote sensing space missions.

The first factor, time, has enabled decades of intense scrutiny of lunar data and materials by a small but dedicated cadre of lunar scientists. This group has explored in great detail the available data and developed the currently known characteristics of the Moon, from its atmosphere to its core. A highly important revelation during this period of contemplation was the formulation of the giant impact hypothesis for lunar origin. The fundamental question of lunar origin persisted beyond Apollo, and prevailing hypotheses all suffered from one or more serious shortcomings. The giant impact hypothesis overcame these problems and meets all known constraints. The new insight was prompted less by working with new data, than by integrating the available data and engaging in deep thinking about the conditions present in the early solar system. The period of contemplation featured many other lesser but still important insights, such as the recognition from detailed crater counts that lunar volcanism extended far later than the youngest dated basalt samples, perhaps being as young as 1 Ga, placing a significant constraint on the lunar thermal history.

A second factor enabling the formulation of the unifying hypotheses has been the exponential improvement in analytical capability that has had a revolutionary impact on the value of lunar samples collected by the Apollo program and Russian Luna missions (1959-1976). These samples have been conserved and very well documented by NASA's curatorial facility, providing a significant body of data, some of which is still untapped. Analysis at spatial scales and analytical precision inconceivable in 1970—especially more recently developed isotopic systems such as Sm-Nd and Hf-W as well as analysis at the nanoscale by secondary-ion mass spectrometry, transmission electron microscopy, and other methods—has produced new insights in the formation of the Moon, from its core to the regolith. This new capability has enabled the "discovery" (or at least strong inference) about the presence of garnet in the lunar mantle; constraints on the processes attendant on a giant impact origin of the Moon, such as evaporative processes in a silicate vapor cloud; precise refinement of the chronological relationships among ancient lunar rocks; and the recognition of pervasive nanoscale processes involved in regolith evolution.

The improvement in analytical technology also included the patient application of ground-based astronomy. High-performance ground-based telescopic remote sensing, especially infrared spectroscopy, shows that the diversity of the lunar crust revealed at millimeter scale in the samples exists at the kilometer scale, and that rock types unknown in the sample collection exist far from the Apollo landing sites. Ground-based radar has revealed new insights into the nature of the lunar regolith and placed tight constraints on the nature of the permanently shaded portions of the Moon. Astronomical observations also enabled the detection of a tenuous atmosphere of sodium and potassium to supplement the constituents discerned by Apollo surface experiments.

Regarding the third factor, the increase in computational capability, a major beneficiary of this new capability has been the giant impact hypothesis. The ability to apply successful computational tests of this hypothesis using two- and three-dimensional fluid dynamics codes graduated what might have been simply an interesting and competitive notion to the status of a ruling paradigm. Another beneficiary of this new capability is the cataclysm hypothesis according to which models of the time evolution of the large bodies in the outer solar system suggest that planet migration could have instigated the cataclysm. Other fields have benefited as well, such as the reanalysis of Apollo seismic data using modern and computationally intensive techniques and analysis and the integration of remote sensing and new spacecraft geophysical data.

With respect to the fourth factor, the recognition of meteorites from the Moon, many of these meteorites have characteristics which suggest that they originate far from the Apollo zone, being extremely poor in incompatible elements—known collectively by the acronym KREEP (for enrichment of some of the first recognized incompatibles: potassium [K], rare-earth elements [REE], and phosphorus [P])—that are characteristic of the Apollo samples. These meteorites raise questions about the notion that the Moon can be understood solely in terms of the samples from Apollo, since the meteorites exhibit subtle differences that may ultimately challenge one or more of the prevailing hypotheses and lend additional impetus to the need for new lunar samples.

The fifth key factor enabling the formulation of the unifying hypotheses is the three post-Apollo missions that returned global lunar remote sensing data. The pioneer was Galileo, which on its way to Jupiter flew by the Moon twice, in 1990 and 1992; its multispectral image data first drove home the compositional distinctiveness of the vast South Pole-Aitken Basin. Next, in 1994, was the Clementine polar lunar orbiter, a joint NASA/Department of Defense technology demonstration mission run in part by lunar enthusiasts who allowed aspects of its sensor payload to be tailored to lunar issues. That 2-month mission resulted in the first unified global remote sensing data set; revealed the large-scale topography of the Moon, including the extent and depth of the South Pole-Aitken Basin; allowed near-global estimates of crustal thickness; provided the first global evaluation of selected elemental and mineral abundances; and brought the importance of the lunar poles into sharp focus with innovative, if controversial, measurements of polar radar properties aimed at the detection of water ice. The third and arguably most scientifically significant mission was the NASA Discovery mission, Lunar Prospector, in 1998. This mission initiated a possible paradigm shift, as it showed the unanticipated extent to which the Moon is asymmetric in composition, in particular for the heat-producing elements. The profound asymmetry, well out of the scope of the lunar magma ocean hypothesis as originally formulated, is prompting a rush of new research attempting to accommodate this fundamental observation into current understanding. Lunar Prospector also identified a concentration of hydrogen at both poles, renewing the discussion of possible volatile deposits.

The more than 30 years of study since Apollo, the technical advances, and the space missions bring us well prepared, at this dawn of a new era of return to the Moon with robots and humans, to test current, durable models and to pose new questions not previously considered. With time, some of these hypotheses have attained the status of "paradigms." Because of the tendency to view a paradigm as accepted truth, it is important to mount challenges constantly, and the new era of lunar exploration provides the opportunity to test each paradigm to varying degrees. Among the highest priorities for scientific understanding will be investigations that either challenge or expand the ruling paradigms.

TESTING THE PARADIGMS

Perhaps the hypothesis least likely to be significantly altered in the near future is the giant impact hypothesis. Tests of this model rest on new samples and on developing a much more detailed understanding of the composition of both Earth and the Moon. Clearly, such a quest will be significantly aided by the receipt of lunar samples from diverse locations and by a more thorough and detailed robotic assessment of the compositional character and structure of the Moon. Despite the directed quest for "genesis rocks," there are no samples as old as the Moon, and the first 100 Ma of lunar history remains obscure. Improvement of the knowledge of the Moon's composition relative to the composition of Earth is also ultimately limited by uncertainties in the composition of the accessible and well-sampled Earth. Anomalies in several isotope systems, including Hf, Nd, and Ba, have recently challenged faith in the chondritic composition of the Earth-Moon system, leading to proposals of unsampled domains on Earth (the core-mantle boundary) and on the Moon. At present, straightforward compelling tests do not exist; however, illumination of the giant impact hypothesis may be possible with new sampling, more abundant data, better analytical techniques, or conceptual breakthroughs.

The lunar magma ocean hypothesis, formulated almost immediately after the receipt of the Apollo 11 samples, was a vital conceptual breakthrough in understanding the early formation of planetary crusts. The concept has provided a new framework for understanding the early Earth, Mars, Mercury, and differentiated asteroids such as Vesta. It explains the observed ancient plagioclase-rich lunar rocks and their relationship to the source area of basalts that cover portions of the Moon through differentiation of a global magma body, the lunar magma ocean (see Figure 2.4). However, the hypothesis is now known to be inadequate to capture everything that has been learned about the early Moon. As Newtonian physics is to Einsteinian physics, the lunar magma ocean hypothesis as formulated in 1970 and refined in the following decades is not wrong, but incomplete; its shortcomings lie not in failures, but in scope. In its pure form, the lunar magma ocean hypothesis is fundamentally a one-dimensional

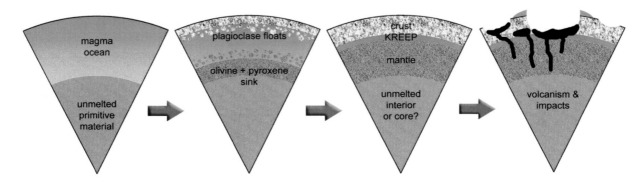

FIGURE 2.4 The lunar magma ocean concept. NOTE: KREEP is defined in Appendix B. SOURCE: From Barbara Cohen, University of New Mexico, adapted from many sources.

crystallization sequence, and a one-dimensional model was all that was required at the close of the Apollo program, when little data existed that placed ancient rocks into a global, three-dimensional spatial context.

Global remote sensing data from the post-Apollo missions forced rethinking of the simple lunar magma ocean hypothesis, principally through their documenting of the distribution and heterogeneity of crustal components. A major enigma is the dramatic asymmetric distribution of thorium, a significant heat-producing element. According to the lunar magma ocean hypothesis, as the globe-encircling magma body cooled and minerals precipitated, a lunar mantle formed from dense olivine and pyroxene, and a crust accumulated composed of buoyant plagioclase. In the waning stages of crystallization, elements that are not accommodated easily into the structures of these major minerals became enriched in the residual melt. This process led to a layer sandwiched between the aluminous crust and the iron- and magnesium-rich mantle that is strongly enriched in incompatible elements, known collectively as KREEP. One of these incompatible elements, thorium, is relatively abundant and was readily detected by the Lunar Prospector's gamma-ray spectrometer (see Figure 2.5).

Apollo measurements from equatorial orbit hinted at a hemispherical asymmetry in Th; Lunar Prospector showed that the thorium (and the rare-earth elements Sm and Nd) and hence KREEP is distributed strikingly

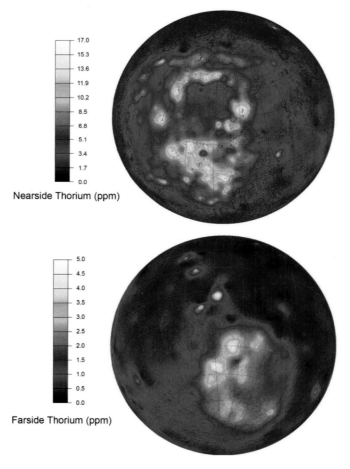

FIGURE 2.5 Spatially deconvolved global thorium abundances measured with the Lunar Prospector gamma-ray spectrometer for the nearside (top) and farside (bottom) of the Moon. Note that the abundance scale bars are different for the nearside and farside. SOURCE: Courtesy of Tom Prettyman and David Lawrence, Los Alamos National Laboratory. Data from D.J. Lawrence, R.C. Puetter, R.C. Elphic, W.C. Feldman, J.J. Hagerty, T.H. Prettyman, and P.D. Spudis, Global spatial deconvolution of lunar prospector Th abundances, *Geophys. Res. Lett.* 34:L03201, 10.1029/2006GL028530, 2007.

asymmetrically and is strongly concentrated on the lunar nearside, with profound implications for lunar thermal evolution. While the lunar magma ocean hypothesis predicted a global zone of incompatibles concentrated at the base of the crust, Lunar Prospector showed that not even the immense farside South Pole-Aitken Basin exposed such a subcrustal enriched layer comparable to that seen in the area sampled by Apollo. High concentrations of Th (and by inference KREEP) are only found on the nearside. The seismic discontinuity provisionally detected at ~500 km depth by Apollo has often been suggested as the base of the original planetwide magma ocean. However, this feature (whose existence is in some doubt) has only been inferred beneath a limited area of the Moon, owing to the restricted geographic span of the Apollo seismic network. An alternative explanation is that it represents the regional depth of melting for the nearside mare basalts and thus might also be consistent with major lateral crustal heterogeneity. If the magma ocean originally was well mixed and essentially one-dimensional, at some epoch its character inflated to three dimensions and the Th (and KREEP) underwent massive lateral migration.

This recent recognition has spurred ongoing efforts to understand the asymmetry within the context of the lunar magma ocean model. Petrologic models suggest that throughout the crystallization of the magma ocean, precipitating minerals became increasingly dense as they accommodated more iron relative to magnesium, with the dense iron-titanium oxide ilmenite being among the last to crystallize (though prior to the crystallization of the last KREEP-rich residuum), perched atop the mantle. This sequence leads to a lunar mantle that was gravitationally unstable, and energetic overturn is enabled. The wide variation in Ti observed in lunar mare basalts in samples and in remote sensing data indicates a highly heterogeneous mantle with respect to Ti and has been cited as evidence of redistribution of magma ocean components. There have been recent attempts to produce models that exploit this gravitational instability to result in large-scale lateral motion. At this early stage, the efforts to model mantle overturn require sensitive and unsatisfying adjustment of viscosity in order to enable this type of global overturn while also preventing breakup into small convection cells; thus, at present this fundamental problem is far from solved, but it is certain that the magma ocean model as it existed prior to Lunar Prospector will require substantial revision and enhancement to take into account the new constraints.

Of the above-named hypotheses, the most vulnerable, because of the existence of well-defined tests aimed at its overturn, is the terminal cataclysm hypothesis. The validity of this hypothesis is the subject of vigorous current debate, and the possible existence of the hypothesized pulse of impacts of large objects has both profound scientific consequences and compelling tests. The consequences derive from the implications of such a pulse for the origin and development of life very early in Earth history and from its implications for the early evolution of the solar system. Astrobiologists have coined the term "impact frustration" to describe the possible effect on terrestrial life of impact by objects hundreds of kilometers in diameter. Models suggest that an impact on Earth comparable with the impact that formed the lunar Imbrium basin would vaporize the oceans and sterilize the top of the crust to depths of hundreds of meters. The hyperthermophilia of Archaea and bacteria near the base of the tree of life suggest a genetic bottleneck only survived by hardy organisms resistant to high temperature. Large objects did strike the Moon, as demonstrated by the many documented impact basins, and therefore a proportionate number also struck Earth. If these impact events occurred over the period 4.5 billion to 3.85 billion years ago, tens to hundreds of millions of years would separate the impacts, enabling the cooling and recovery of existing life and the population of refuges against the next impact. However, a narrow pulse of dozens of such impacts in a few million years would cause greater heating and leave little time for recovery and repopulation of refuges between events. This leads to the suggestion that any surface life that persists today must have either originated after the cataclysm (i.e., after about ~3.85 Ga, which appears to be the time of the last of the lunar basins) or survived deep in the crust.

What are the possible implications? This late date of heavy bombardment in the Earth-Moon system would allow half the time believed necessary to produce the extreme biochemical complexity observed in the Archaea and bacteria. It would also allow only half the time likely necessary for the organisms that produced the first generally accepted microfossils at 3.5 Ga. This implies that the origin of terrestrial life and its development to the level of biochemically complex and sophisticated microorganisms occurred at an exceedingly fast rate, over only a few hundred millions years at most. If the intensity of late bombardment was heavy enough to sterilize the surface of Earth at ~3.9 Ga, then it is possible that life emerged more than once on Earth, but that all present life is derived from the last developing life, not the first life on Earth. On the other hand, if the observed lunar impact basins formed over a much longer period (with no peaked cataclysm occurring), the development of life was permitted to occur at a more leisurely pace, and planetary recovery from each impact event was sustained by hundreds of millions of years of relative quiet.

Models to explain the apparent lunar cataclysm also have long-reaching implications for our understanding of solar system dynamics. For instance, one such model postulates that exchange of angular momentum between the giant planets and an early solar system teeming with smaller objects causes the orbits of the large planets to migrate. As a consequence, the giant planets pass through highly disruptive resonances that ought to scatter small objects formed in the outer solar system both outward and inward. The excited orbits of the Kuiper Belt objects are testimony to a scattering event or events that may also have sprayed the inner solar system with icy objects large and small long after planetary accretion had waned and background impact fluxes had drastically dropped. These planet migration and scattering calculations solve profound problems with the terminal cataclysm, that is, where were the objects stored, and what precipitated their bombardment of the inner solar system? They also suggest that Mars too would have experienced a cataclysm, removing that planet as a safe refuge for life to later populate Earth by way of meteorite exchange between planets.

Work on early Earth is beginning to intersect work on the early Moon and to provide critical constraints on both bodies. In particular, the implications of early bombardment and the terminal cataclysm hypothesis must also accommodate new results from the study of early Earth which strongly suggest that relatively clement conditions existed on Earth as early as 4.4 Ga to 4.2 Ga, raising the possibility that life might have emerged within 200 Ma of the formation of Earth (see Figure 2.6).

FIGURE 2.6 "First Lunar Tides," an artist's conception of the Moon as seen from Earth about 4.2 billion years ago. The heavily cratered Moon is in close Earth orbit, and mare basalts have not yet filled the crater bottoms. The oceans have recently condensed on the cooling surface of Earth and are experiencing the first tides. It is not known if life existed in these early seas. Likewise, the size and frequency of impact events are uncertain, as is the effect of these events on the emergence or extinction of life. The ravages of tectonics will destroy most evidence of this time on Earth, but the lunar surface remains. SOURCE: Don Dixon, with permission.

Zircons (which are incredibly difficult to alter) with ages as great as 4.4 Ga have been found in the ancient terrestrial Archaean metasediments (metamorphosed but recognizably sedimentary rocks). Because most terrestrial zircons are formed in granitic rocks, these ancient zircons imply the presence of continent-building activity at the time of zircon formation, rather than the presence of a simple basaltic crust. The mildly elevated values of $\delta^{18}O$ in these zircons support the proposal that, during this period, chemical weathering and erosion had occurred, both of which require liquid water to alter the protoliths (original rocks) of subsequently formed granitic rocks. If the high impact flux suggested by lunar chronology could have caused the oceans to vaporize repeatedly, there may be an inconsistency in the lunar and terrestrial observations. More investigations of the lunar record and terrestrial history are indicated.

As on the Moon, terrestrial zircons are the end product of extensive igneous processing, and they host some of the important incompatible elements of the KREEP association (of potassium, rare-earth elements, and phosphorus found in lunar basalts). Zircons generally require an evolved host rock such as granite, and on Earth the presence of large-scale processes giving rise to granite and mountain belts account for most, though not all, zircons. In addition to hosting a range of incompatible elements, zircons are extremely durable and are resistant to melting, abrasion, and weathering. It is possible that these minerals are the end result of differentiation of basaltic magma, but their sheer abundance in Archaean sedimentary rocks on Earth suggests that most were derived from rocks of evolved, possibly granitic, composition. Radiometric dating of these detrital zircons shows that some were formed before 4.0 Ga, with several terrestrial ages near 4.4 Ga. Elevated oxygen isotope ratios in the most ancient zircons appear to require that the protoliths were altered by liquid water as long ago as 4.3 Ga. Although work on these zircons has only begun, they already indicate the presence of buoyant, probably granitic components in the crust shortly after the formation of Earth (while the lunar magma ocean completed crystallizing many of its primary rocks), and the zircon host rock was profoundly affected by the presence of liquid water. Further study of ancient zircons and rocks from Earth may resolve these differences. Relatively few zircons have been studied from the Moon, and further analysis will elucidate differences in magmatic environment. It is also possible that zircons from otherwise-unsampled domains of early Earth will be located on the Moon, delivered as meteorites, if sufficient quantities of regolith are searched.

The terminal cataclysm hypothesis can be definitively tested by measuring the ages of large impact basins that are far from the Apollo sampled zone. The basis for the hypothesis, a cluster of radiometric dates near 3.8 Ga, may all be related to the vast Imbrium basin that dominates all portions of the Moon visited by Apollo. Thus, sample-derived radiometric dates of basins far from Imbrium, and most especially the largest and oldest basin of them all, South Pole-Aitken Basin, will rigorously test the terminal cataclysm hypothesis.

THE LUNAR ATMOSPHERE

The lunar atmosphere links the Sun and solar system volatiles through the lunar surface. The Sun provides one input, the solar wind, to the lunar environment, some of which is entrained in the lunar plasma environment, some of which is implanted into grains of the lunar surface, and some of which serves as a critical loss mechanism for the lunar neutral atmosphere. Hydrogen and helium are the dominant species that saturate the outermost surfaces of lunar soil grains, but the solar wind provides a wide diversity of other components, particularly carbon and nitrogen. Many other sources can provide volatile inputs to the lunar environment: comets, asteroidal meteorites, interplanetary dust particles, the transit of interstellar giant molecular clouds, Earth itself (ions from its atmosphere are continuously shed down the magnetotail or episodically removed as impacts eject atmosphere into space), and even the Moon's interior (which may occasionally still outgas at high rates, as suggested by some young geologic features). These sources provide volatile species that may then be transported across the lunar surface until either lost or sequestered in lunar sinks. This atmosphere is extremely tenuous, having a total mass of a few tons. Research shows that the lunar atmosphere is generally resilient and can restore its original state even a few weeks or months after a disturbance, but its present state is relatively fragile against extended large-scale lunar activities.

The lunar atmosphere may actually be dominated by dust, although its properties are not well known. The mass of suspended dust may be larger than the mass of the atmosphere. Observations by the Surveyor, Apollo, and Clementine missions showed the presence of high-altitude dust, presumably lofted by electrostatic levitation.

Permanently shaded regions of the lunar poles may constitute one of the important sinks for volatiles. Lunar Prospector showed excesses of hydrogen at the 50 km to 100 km scale near the poles, although radar detects no widespread thick deposits of volatiles (unlike at the poles of Mercury). Despite the fact that these cold, shadowed surfaces are expected to act as lunar cold traps, their effectiveness in trapping volatiles for long periods is unknown. Regardless of total efficiency, the lunar polar soil may contain a wealth of scientific information on solar system volatiles.

Significant gaps remain in the understanding of the lunar atmosphere system, especially the three-dimensional flows and nature of constituents through the atmosphere, the detailed behavior of dust and its relationship to the vapor component, and the role and state of the polar cold traps. It is safe to say that the dynamic system exists, but there is no predictive understanding of its behavior.

The more than 30 years since the Apollo program have enabled lunar and other scientists to make great strides in the understanding of the origin and evolution of the Moon, Earth, and other planets using the Moon as a lens. But this era has also revealed troubling defects in that lens owing to the lack of key data in time, space, and physical characteristics. Building on the foundation of the Apollo program and over 30 years of contemplation and technical progress, a well-formulated new era of lunar exploration will enable key tests of hypotheses major and minor that may revolutionize the understanding of the origin and evolution of the Moon, Earth, and the planets.

3

Science Concepts and Goals

SCIENCE CONCEPTS AND KEY SCIENCE GOALS IDENTIFIED WITH EACH:
A TABULAR PRESENTATION

This chapter discusses eight specific areas of scientific research, termed science concepts, that can and should be addressed during implementation of the initial phases of the Vision for Space Exploration (VSE) as well as the early phases of human exploration activities. An integrated set of prioritized science goals is presented with each concept. Opportunities for the science of Earth and the universe from the Moon are discussed separately in Chapter 6. The science concepts are listed in priority order in Table 3.1 in terms of scientific value. A full discussion of the prioritization of these science concepts and the science goals is contained in Chapter 5.

The key science goals identified with each science concept are summarized in Table 3.1 and discussed extensively in this chapter. The overarching themes of lunar science presented in Chapter 1 are not tied to a single endeavor, but emerge from a multitude of such goals and concepts as broad topics of fundamental science required in order to attain understanding of the solar system. Since the origin and evolution of life in the universe are also of great interest to scientists and the public, those goals that may have important implications for life are highlighted separately in Table 3.1.

DISCUSSION OF SCIENCE CONCEPTS AND KEY SCIENCE GOALS IDENTIFIED WITH EACH

Concept 1: The bombardment history of the inner solar system is uniquely revealed on the Moon.

The heavily cratered surface of the Moon testifies to the importance of impact events in the evolution of terrestrial planets and satellites and the exceptional ability of the lunar surface to record them all. Lunar bombardment history is intimately and uniquely intertwined with that of Earth, where the role of early intense impacts and the possible periodicity of large impact events in the recent past on the atmosphere, environment, and early life underpin our understanding of habitability. The correlation between surface crater density and radiometric age discovered on the Moon serves as the basis for estimating surface ages on other solid bodies, particularly Mars. Significant uncertainties remain in the understanding of the lunar cratering record and our ability to extend it to other planets of the solar system. Returning to the Moon provides an unparalleled opportunity to resolve some of the fundamental questions that relate to these goals.

TABLE 3.1 Primary Science Goals of Lunar Science Concepts and Links to Overarching Themes

Science Concepts	Science Goals	Early Earth-Moon System	Terrestrial Planet Differentiation and Evolution	Solar System Impact Record	Lunar Environment	Implications for Life
		Overarching Themes				
1. The bombardment history of the inner solar system is uniquely revealed on the Moon.	1a. Test the cataclysm hypothesis by determining the spacing in time of the creation of lunar basins.	X		X		X
	1b. Anchor the early Earth-Moon impact flux curve by determining the age of the oldest lunar basin (South Pole-Aitken Basin).	X	X	X		X
	1c. Establish a precise absolute chronology.	X	X	X		X
	1d. Assess the recent impact flux.			X	X	X
	1e. Study the role of secondary impact craters on crater counts.			X		
2. The structure and composition of the lunar interior provide fundamental information on the evolution of a differentiated planetary body.	2a. Determine the thickness of the lunar crust (upper and lower) and characterize its lateral variability on regional and global scales.		X	X		
	2b. Characterize the chemical/physical stratification in the mantle, particularly the nature of the putative 500-km discontinuity and the composition of the lower mantle.		X			
	2c. Determine the size, composition, and state (solid/liquid) of the core of the Moon.	X	X			
	2d. Characterize the thermal state of the interior and elucidate the workings of the planetary heat engine.	X	X			
3. Key planetary processes are manifested in the diversity of lunar crustal rocks.	3a. Determine the extent and composition of the primary feldspathic crust, KREEP layer, and other products of planetary differentiation.		X			
	3b. Inventory the variety, age, distribution, and origin of lunar rock types.	X	X		X	
	3c. Determine the composition of the lower crust and bulk Moon.	X	X			
	3d. Quantify the local and regional complexity of the current lunar crust.		X	X		
	3e. Determine the vertical extent and structure of the megaregolith.			X	X	X
4. The lunar poles are special environments that may bear witness to the volatile flux over the latter part of solar system history.	4a. Determine the compositional state (elemental, isotopic, mineralogic) and compositional distribution (lateral and depth) of the volatile component in lunar polar regions.				X	X
	4b. Determine the source(s) for lunar polar volatiles.			X	X	
	4c. Understand the transport, retention, alteration, and loss processes that operate on volatile materials at permanently shaded lunar regions.				X	
	4d. Understand the physical properties of the extremely cold (and possibly volatile rich) polar regolith.				X	
	4e. Determine what the cold polar regolith reveals about the ancient solar environment.				X	

(continued on next page)

TABLE 3.1 continued

Science Concepts	Science Goals	Overarching Themes					Implications for Life
		Early Earth-Moon System	Terrestrial Planet Differentiation and Evolution	Solar System Impact Record	Lunar Environment		
5. Lunar volcanism provides a window into the thermal and compositional evolution of the moon.	5a. Determine the origin and variability of lunar basalts.		X				
	5b. Determine the age of the youngest and oldest mare basalts.		X	X			
	5c. Determine the compositional range and extent of lunar pyroclastic deposits.		X		X		
	5d. Determine the flux of lunar volcanism and its evolution through space and time.		X	X	X		
6. The Moon is an accessible laboratory for studying the impact process on planetary scales.	6a. Characterize the existence and extent of melt sheet differentiation.	X	X	X			
	6b. Determine the structure of multi-ring impact basins.	X		X			
	6c. Quantify the effects of planetary characteristics (composition, density, impact velocities) on crater formation and morphology.		X	X			X
	6d. Measure the extent of lateral and vertical mixing of local and ejecta material.			X	X		
7. The Moon is a natural laboratory for regolith processes and weathering on anhydrous airless bodies.	7a. Search for and characterize ancient regolith.				X		
	7b. Determine physical properties of the regolith at diverse locations of expected human activity.				X		
	7c. Understand regolith modification processes (including space weathering), particularly deposition of volatile materials.				X		
	7d. Separate and study rare materials in the lunar regolith.		X	X	X		X
8. Processes involved with the atmosphere and dust environment of the moon are accessible for scientific study while the environment remains in a pristine state.	8a. Determine the global density, composition, and time variability of the fragile lunar atmosphere before it is perturbed by further human activity.				X		
	8b. Determine the size, charge, and spatial distribution of electrostatically transported dust grains and assess their likely effects on lunar exploration and lunar-based astronomy.				X		
	8c. Use the time-variable release rate of atmospheric species such as ^{40}Ar and radon to learn more about the inner workings of the lunar interior.		X		X		
	8d. Learn how water vapor and other volatiles are released from the lunar surface and migrate to the poles where they are adsorbed in polar cold traps.				X		

Science Goal 1a—Test the cataclysm hypothesis by determining the spacing in time of the creation of lunar basins.

From the returned samples and crater statistics, it is known that that during the past 3 billion years (Ga) the lunar impactor flux was relatively constant with possible variations by a factor of two, which is in good agreement with age data for young impact-melt rocks. It is also known that before 3 Ga ago, the impactor flux was much higher and rapidly decayed in time. The lunar chronology curve is well constrained, with small errors in the age range from about 4.0 Ga to 3.0 Ga. However, major uncertainties still exist for the pre-Nectarian period (more than about 4 Ga) and for the Eratosthenian and Copernican periods (less than about 3 Ga). The steepness of the calibration curve at ages older than about 3.75 Ga, the possibility that the pre-Nectarian surfaces for which crater counts exist may not be older than 4.2 Ga, and the fact that impact-melt lithologies older than 4.15 Ga are lacking indicate that the cratering rate may not increase smoothly according to the present calibration curve from 3.75 Ga back to the time of the formation of the Moon. The radiometric ages of impact-melt lithologies in the returned-sample collection have been used as an argument for a late cataclysm, that is, a spike in the cratering rate around 3.9 Ga ago, just about the time the life on Earth was emerging. However, a possible explanation of the spike is that the geologic setting of the Apollo landing sites led to a selective sampling of material from Nectaris, Serenitatis, and Imbrium. Most observers now agree that the gas retention ages of meteorites from the asteroid belt do not show the spike, but rather only a modest, broad peak from ~4.2 Ga to ~3.5 Ga ago. Furthermore, dated impact-melt clasts in Th-poor, KREEP-poor lunar meteorites, which most likely sample other areas of the Moon besides the Th-rich, KREEP-rich regions among Apollo sites, do not show evidence for a cataclysm at 3.9 Ga (KREEP is the acronym for potassium [K], rare-earth elements [REE], and phosphorus [P]). With currently available data, it is impossible to decide whether a cataclysm occurred or whether the cratering rate smoothly declined with time after lunar origin. This controversy is extremely important because it affects not only lunar science, but the understanding of the entire solar system. For example, models of planet formation and accretion of planetary debris are often based on the supposition that cataclysmic formation of all lunar basins between about 4.0 Ga and 3.8 Ga ago is well established. Determining the ages of impact-melt rocks in lunar meteorites and/or from the South Pole-Aitken (SPA) Basin (the stratigraphically oldest lunar basin; see Figure 3.1) and major impact basins within the SPA Basin will probably resolve this issue. The precision required to date these events accurately requires isotopic analysis of well-chosen samples in terrestrial laboratories.

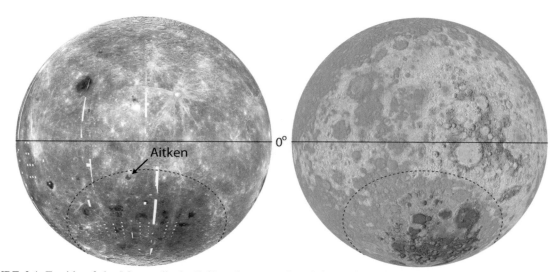

FIGURE 3.1 Farside of the Moon: albedo (left) and topography (right) derived from Clementine data. The outer ring of South Pole-Aitken (SPA) Basin is shown with a dotted line (after Wilhelms, 1987). The enormous SPA Basin is not located at the South Pole, but derives its name from the fact that it extends from the South Pole to the crater Aitken near the equator. SOURCE: Courtesy of Noah Petro and Peter Isaacson, Brown University.

Science Goal 1b—Anchor the early Earth-Moon impact flux curve by determining the age of the oldest lunar basin (South Pole-Aitken Basin).

The pre-Nectarian period is the time span beginning with the formation of the Moon and ending with the impact of the Nectaris basin, which occurred at ~3.92 Ga to ~4.1 Ga. Since the oldest age of solid lunar surface material determined so far is 4.52 Ga, the pre-Nectarian period is ~600 million years (Ma) long. Our knowledge of the pre-Nectarian system stems from the photogeological identification of some 30 multi-ring basins, including the oldest known basin, the South Pole-Aitken Basin, and their ejecta deposits and returned samples of rocks with absolute ages older than Nectaris. Unfortunately, it is impossible to directly relate the dated pre-Nectarian rock clasts to any specific pre-Nectarian geologic surface unit, because subsequent multiple impact events displaced these samples after their formation. The relative ages of most of the pre-Nectarian multi-ring basins are based on crater counts on their ejecta formations. On the basis of these crater counts, it has been suggested that no multi-ring basins older than 4.2 Ga are unequivocally recorded. This implies that the oldest basins, including the South Pole-Aitken and Procellarum basins, were formed between 4.2 Ga and 4.1 Ga. However, these ages are poorly constrained because the precise age of the oldest basin, namely, the South Pole-Aitken Basin, has not yet been determined by isotope dating of returned samples. As the oldest multi-ring impact basins are important calibration points for the lunar chronology curve, this has implications with respect to the knowledge of the exact shape of the cratering chronology, which is the basis for absolute dating not only of the lunar surface, but also of other planetary surfaces. Therefore, it is paramount to date the oldest impact basin precisely in order to derive the exact shape of the lunar chronology.

Science Goal 1c—Establish a precise absolute chronology.

While most geologists agree on absolute ages of the different terrestrial chronostratigraphical systems, there is debate on the ages of the lunar systems. Traditionally the lunar chronostratigraphic systems are based on ejecta blankets of large impact craters and basins that serve as marker horizons similar to terrestrial fossil or ash layers. These impact basins and young craters are stratigraphic benchmarks that allow the global determination of the relative age of lunar surfaces that have not been or cannot be directly accessed. However, there is still considerable debate about the ages of individual impact basins and craters on the Moon. At least two chronostratigraphical systems were defined by the existence or absence of bright rays. Eratosthenian craters would lack such rays, whereas Copernican craters would show these rays. However, newer studies indicate that age assignments solely based on rays might not be reliable. In fact, samples from Apollo 15 that are interpreted to represent material from the rayed crater Autolycus are 2.1 Ga old, hence Eratosthenian in age. Based on returned Apollo 12 samples, which were collected on one of the rays of Copernicus crater, the crater formed about 800 Ma to 850 Ma ago. While radiometric ages of Apollo 12 samples suggest a narrowly constrained age of 800 Ma to 850 Ma for Copernicus, crater counts on the ejecta blanket of Copernicus indicate a significantly older age, of up to 1.5 Ga. This could mean that material from Copernicus was not collected or that the samples do not represent the surface material dated with crater counts. The timing of Tycho was inferred from a landslide on the slopes of the South Massif and the "Central Cluster" craters at the Apollo 17 landing site, interpreted as secondary craters from Tycho. Based on this interpretation, an age of ~100 Ma was proposed for Tycho. However, the geological evidence for the South Massif landslide and the Central Cluster craters being formed by distant ejecta from Tycho remains somewhat equivocal. The exact ages of Copernicus and Tycho are important because they provide important calibration points for the lunar chronology at young ages.

There is also debate on the beginning of the Late Imbrian period, which is defined by the Orientale impact, with some authors favoring an age of 3.72 Ga and others favoring an age of 3.75 Ga. However, it could be almost as old as Imbrium, that is, 3.84 Ga. The problem is that the Orientale basin cannot be dated precisely because no samples in the current collection can unambiguously be attributed to the Orientale formation event. Similarly, there are large uncertainties associated with the beginning of the Early Imbrian period, with ages ranging from 3.77 Ga to 3.91 Ga, depending on which ages are used for the Imbrium impact itself. Finally, the base of the Nectarian period could be as old as 3.85 Ga, 3.92 Ga, or 4.1 Ga, depending on the interpretation of radiometrically dated samples.

Science Goal 1d—Assess the recent impact flux.

Variability in the recent lunar and terrestrial impact flux may be related to singular solar system dynamic events, such as asteroid breakups, and may have significant effects such as impact-induced mass extinction on Earth (e.g., the Cretaceous/Tertiary [K/T] boundary and currently hotly debated Permian/Triassic [P/Tr] boundary events). On Earth, there is a greater relative number of younger craters, possibly suggesting a recent increase in projectile flux but more likely related to the difficulty of preserving and accessing older craters on a surface shaped by erosion, deposition, and plate tectonics. The Moon more faithfully records the projectile flux in the Earth-Moon system over the past ~3.5 Ga, and researchers can use it to determine whether this flux has been approximately constant, or has exhibited shorter-term variations or periodicity, by determining crater densities on known young surfaces, such as crater ejecta deposits, and radiometric ages of lunar soil spherules. Current literature is contradictory as to whether the impact rate in the inner solar system has increased by a factor of two, stayed constant, or decreased by a factor of three in the last 3 Ga. Since the gross long-term time behavior must have been the same throughout the inner solar system, and perhaps in the outer solar system, radiometric dating of a large number of randomly selected primary impact craters would thus refine the cratering chronology system for all the planets. With more than 35 years passed since the first high-resolution images were taken, it is also possible to directly study how many impact craters have formed since then in order to derive precise estimates of the recent impact flux.

Science Goal 1e—Study the role of secondary impact craters on crater counts.

New studies of the martian impact crater Zunil revealed a surprisingly large number of small secondary impact craters. Consequently, it was argued that a similarly large number of small impact craters on the Moon could be secondaries. If true, the flux of small primary impact craters on the Moon might have been overestimated, which could have effects on the precise shape of the standard distribution. However, other groups argued that Zunil might only be a special case, not representative for lunar impact craters in general, and that secondary craters can easily be detected and omitted from crater counts. Detailed studies of young lunar impact craters and the distribution and number of their secondary impact craters will help to better understand the process of secondary impact cratering and its possible effects on crater statistics.

Other factors may affect crater counts. Recently there have been presented models on latitudinal asymmetries of up to 20 percent in the number of formed impact craters. Such asymmetries would be greatest for planets that do not have large variations in obliquity, such as Mercury. For this reason it is important to test these models before extrapolating the lunar chronology to other planets. Similarly, it has been argued that there are differences in the cratering rate between the leading and trailing sides of planetary bodies. The degree of these asymmetries depends on the velocities of the celestial projectiles and the Earth-Moon distance.

In summary, the overarching science requirement is to characterize and date the impact flux (early and recent) of the inner solar system.

Concept 2: The structure and composition of the lunar interior provide fundamental information on the evolution of a differentiated planetary body.

One of the key motivations for studying the Moon is to better understand the origin of the planets of the inner solar system in general and that of Earth in particular. The origin of the Moon is inextricably linked to that of Earth. The precise mode of formation affected the early thermal state of both bodies and, therefore, affected the subsequent geologic evolution. The leading hypothesis at present is that the Moon formed as the result of the impact of a Mars-sized object with growing Earth. However, the details of the process are not clear. Because the Moon's geologic engine largely shut down long ago, its deep interior is a vault containing a treasure-trove of information about its initial composition, differentiation, crustal formation, and subsequent magmatic evolution. During and immediately after accretion, the Moon underwent primary differentiation involving the formation of a (presumably) iron-rich core, a silicate mantle, and a light, primordial crust. The initial bulk composition, as well as the pressure and temperature conditions during this separation, will be reflected in its current chemistry, structure, and dynamics. Although researchers have some information on the composition of the outermost layers of the Moon's crust, that of the bulk crust is less well known, and even its thickness is not well established. The composition of the mantle can only be vaguely estimated, and the presence of compositional stratification, bearing on the late stages of differentiation and the efficiency of subsequent convective mixing, cannot be confirmed or denied. Furthermore, the size and the composition (e.g., Fe versus FeS, or even $FeTiO_3$) of its core are unknown, except for loose bounds on its diameter.

Science Goal 2a—Determine the thickness of the lunar crust (upper and lower) and characterize its lateral variability on regional and global scales.

Apollo-era analyses of seismic data deduced a mean crustal thickness of around 60 km. However, recent reanalyses of the seismic data indicate that a thinner crust, perhaps 30 km to 45 km thick, is more likely. The amount and quality of available seismic data suitable for constraining the crust are limited, resulting in an uncertainty in crustal volume of nearly a factor of two. There is also an indication of a marked increase in velocity at a depth of about 20 km, which may indicate a more mafic and noritic lower crustal layer with uncertain origin. All of these results come from a single region, and any lateral variability is currently constrained only by non-unique gravity modeling.

Science Goal 2b—Characterize the chemical/physical stratification in the mantle, particularly the nature of the putative 500-km discontinuity and the composition of the lower mantle.

What is the nature and extent of the 500 km seismic discontinuity? Seismic velocity profiles, derived from ray paths that, for the Apollo network, extended primarily beneath the Procellarum KREEP Terrane (PKT), suggest that a major seismic velocity discontinuity occurs approximately 500 km below the surface (although the existence of this discontinuity has been questioned). The magnitude of this velocity increase would imply that the discontinuity is compositional in origin, with the deeper mantle likely more aluminous or Mg-rich. One possibility is that this boundary could represent the maximum depth of the lunar magma ocean, and an Al- and Mg-rich primitive mantle exists below 500 km. In this case all the Al in the primordial lunar crust must have been extracted from the upper mantle, and much of the mare basaltic magmatism must have involved melting above 500 km. However, some petrologic models imply a depth of melting of at least 1,200 km. If the magma ocean was deeper, then this discontinuity might mark the transition between early olivine- and later orthopyroxene-rich cumulates. Finally, it is possible that it is not a global feature of the mantle at all, but instead merely corresponds to the maximum depth of melting of the local PKT mare source. In any case, the presence of either radial stratification of lateral heterogeneity on this scale in the mantle has major implications for convection and mantle mixing since solidification of the magma ocean.

What is the nature of the lower mantle? In simple density-driven models of magma-ocean crystallization, mantle cumulates become increasingly rich in iron with the progress of crystallization. This mantle cumulate pile would have been gravitationally unstable. It is likely that late-stage ilmenite cumulates should have sunk through the mantle, and the deep, olivine-rich mantle cumulates should have participated in the overturn as well, bringing early-crystallized magnesium-rich olivine to the upper mantle. But models disagree as to the timing, extent, and efficiency of this overturn. Did it act to homogenize the composition of the mantle? Was mixing inefficient, retaining any original stratification within the magma-ocean cumulates, perhaps in an inverted sequence? Was the overturn global, or was it localized in particular regions of the mantle? Seismic information from Apollo becomes increasingly vague below 800 km, providing virtually no illumination of the scientifically critical zone of the Moon's interior below ~1,100 km. Is there stratification in the deep mantle, or is it well mixed? Is the lower mantle characterized by an increased proportion of Mg-rich olivine, or is there a significant garnet cumulate? Is the lack of observed seismic activity from the farside indicative of partial melting below 1,150 km? What is the mechanism for generating periodic moonquakes at depths of >600 km? Do the lateral variations seen in the upper crust have roots extending deep into the planet? These questions all bear directly on the nature of the magma ocean and the subsequent thermodynamic history of the Moon.

Science Goal 2c—Determine the size, composition, and state (solid/liquid) of the core of the Moon.

The size, composition, and state of the lunar core are nearly unknown. Yet these parameters have far-reaching implications. Available evidence suggests that the Moon has a small core, less than 460 km radius. It is likely composed of metallic iron (with some amount of alloying Ni, S, and C), although existing geophysical data are also consistent with a core composed of a dense molten Ti-rich silicate magma. The size and composition of the core are fundamentally important for understanding the Moon's origin and evolution. In the giant impact hypothesis, a collision between Earth and a Mars-sized object ejects debris and vapor into circumterrestrial orbit. Most of this debris should come from the silicate mantle of the impactor, but current models do not uniquely constrain the amount of iron that would be entrained in this material. Knowledge of the size and composition of the lunar core could constrain the many unknown parameters associated with these models. The formation of an iron core may have affected the composition of the lunar mantle and subsequently the composition of the mare basalts. An early lunar dynamo within an iron-rich core might help explain the magnetizations that have been measured in some of the Apollo samples and the crustal magnetic fields that have been mapped from orbit.

Science Goal 2d—Characterize the thermal state of the interior and elucidate the workings of the planetary heat engine.

The evolution of the Moon from accretion to its present state is inextricably tied to its thermal history. The amount of heat produced through time and the manner in which it is transported to the surface constitute the engine that drives virtually all geological and geochemical processes. Effective models of these processes require knowledge of fundamental parameters that include interior structure and thermal state. Apollo heat flow measurements are inadequate, both in number and location, to effectively address the current flux of energy from the interior. This leaves a huge gap in knowledge of the bulk composition of the Moon in terms of heat-producing elements, which was a key determinant of the evolutionary path that led to the present state. More generally, the thermal state of the interior, which can be inferred from electromagnetic sounding of the Moon by utilizing its passage through Earth's magnetosphere and perhaps from the physical state of the core (solid versus molten), provides further insight into areas such as the radial partitioning of radiogenic elements. The existence of an early dynamo, which might leave evidence in the remanent magnetization of igneous rocks, would also constrain the thermal history through inferences of the timing of formation and maintenance of a conducting core.

The Moon does not at present have an active core dynamo, but it has numerous localized remanent crustal magnetic regions of around a few kilometer to hundreds of kilometer scale distributed over its surface, indicating the presence of strong magnetizing fields in the past. Measurements of remanent magnetism on Earth provided the crucial evidence for the understanding of the evolution of Earth's interior and surface (e.g., seafloor spreading and plate tectonics), and understanding the processes responsible for lunar magnetism hold similar promise. Studies of lunar crustal magnetism could provide a powerful tool for probing the thermal evolution of the lunar crust, mantle, and core, as well as the physics of magnetization and demagnetization processes in large basin-forming impacts. It is also possible that determining the distribution and properties of the strong magnetic anomalies first observed by Apollo will clarify potential magnetic shielding benefits for co-located lunar bases. This investigation will require a focused program of high-resolution mapping of crustal magnetic fields from orbit, together with surface magnetometer surveys of select regions and the return of samples whose orientation on the Moon was recorded.

Data concerning interior structure and dynamics are difficult to obtain but are worth considerable effort to do so. Direct samples of mantle rocks, from the deeply excavated South Pole-Aitken Basin or xenoliths, can provide a detailed glimpse into lower-crust and perhaps even upper-mantle composition. Geophysical measurements are often the best and in some cases the only way to obtain information about the composition and structure of the deep lunar crust, mantle, and core. It is well known from terrestrial experience that seismology is the most sensitive tool for determining deep internal structure. The waves produced by seismic events can provide essential information bearing on crust and mantle structure and composition and on the size and nature of the core. These measurements can be augmented with analysis of rotational dynamics from precision tracking of surface reflectors. The flow of heat from the Moon's interior is a primary indicator of the global energy budget in terms of sources (e.g., radiogenic, accretional) and the mechanisms that control its release (convection, conduction, volcanism). Knowledge of the heat flow provides important constraints (through inferences about internal temperatures) on the rheology and dynamic behavior of deeper layers of the Moon, on both global and regional scales. It may also be used to address the depth extent of the asymmetric distribution of KREEP. Magnetic induction studies can probe the conductivity of the deep interior to help constrain the temperature profile as well as aid in the characterization of the core.

Thus, a variety of geophysical and compositional analyses of the Moon will enable researchers to determine the internal structure and composition of a differentiated planetary body.

Concept 3: Key planetary processes are manifested in the diversity of lunar crustal rocks.

Like Earth, the Moon possesses a crust, a mantle, and possibly even a small core. However, the Moon represents a very different end member of planetary evolution than Earth. The current understanding of the evolution of the Moon is framed by the lunar magma ocean, a concept developed on the basis of Apollo sample studies. According to this understanding, after the Moon accreted, it was completely molten to a depth of hundreds of kilometers. As the molten Moon cooled and crystallized, olivine, pyroxene, and other mafic minerals sank to form a mantle, while plagioclase floated to form the crust. The last liquid to crystallize, containing incompatible elements such as potassium, rare-earth elements, and phosphorus (collectively known as KREEP), was sandwiched between the crust and the mantle. Remelting of the mantle cumulate package drove fire-fountaining, lava flows, and plutonic emplacement, while large impact craters excavated and distributed the KREEP layer over the lunar surface. In this very simplified model, the composition of the crust has been thought of as generally homogeneous at any point on the Moon, modified only by thin, surface basalt flows or later impacts that scrambled the upper crust (regolith).

Although the concept of the lunar magma ocean continues to serve scientists well, geophysical, remote sensing, and sample analyses reveal a lunar crust that varies both laterally and vertically in composition, age, and mode of emplacement (Figure 3.2). The traditional, dichotomous mare-highland classification developed from Apollo experience is inadequate in describing the structure and geologic evolution of the lunar crust. From the global remote sensing coverage of the Clementine and Lunar Prospector missions of the 1990s and the study of lunar meteorites, researchers now know that they have an incomplete sampling of the lunar crust. Armed with a more global view of the Moon, scientists can now pose sophisticated questions about the lunar crust that will uniquely further the understanding of differentiation processes and the origin of the Moon.

Science Goal 3a—Determine the extent and composition of the primary feldspathic crust, KREEP layer, and other products of planetary differentiation.

Instead of simply mare and highlands, large regions of the Moon have distinct geologic and geochemical characteristics. The surface expression of thorium and iron reveals swathes of the Moon that stand out from one another. These global terrains are inferred to be the result of asymmetry in the crystallizing lunar magma ocean or later large impact events. Large areas of the Moon are covered by nearly pure anorthosite, now called the Feldspathic Highlands Terrane (FHT). Modifying the feldspathic crust is the largest and deepest impact basin on the Moon, the South Pole-Aitken Basin, which may have penetrated a lower crust or mantle component that makes the floor of the basin more mafic than the pristine feldspathic crust. Smaller basins on the lunar farside do not appear to have penetrated deeper than the feldspathic crust, but on the nearside, where the feldspathic crust is thinner, the KREEP layer underlying the crust was excavated by the large, late impact basins such as Imbrium, creating the incompatible-element-rich area called the Procellarum KREEP Terrane (PKT).

Relating these geochemical terranes to the understanding of lunar formation and differentiation remains a fundamental goal of lunar science and will help guide understanding of products of magma oceans on other planets.

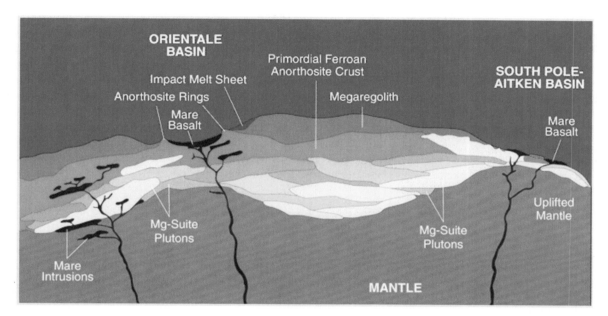

FIGURE 3.2 The complexity of today's lunar crust, showing craters, plutons, magma conduits, and other features, based on a concept by Paul D. Spudis (Applied Physics Laboratory, Johns Hopkins University). The topmost layer of the crust is composed of a mixture of underlying anorthosite (rock containing more than 90 percent plagioclase feldspar) and lower crustal intrusions of Mg-suite magmas. Mg-suite magmas are slightly younger than anorthosites and may have formed when magma became trapped inside the anorthosite crust. This complicated picture is actually simplified from reality, which makes determining the bulk chemical composition of the lunar crust a difficult business. SOURCE: Courtesy of Planetary Science Research Discoveries, University of Hawaii.

Because of its enormous size the South Pole-Aitken Basin is expected to have excavated more deeply than any other visible lunar basin, but it does not contain a significant KREEP component in its ejecta, whereas the nearside basins do. This appears to reflect a fundamental asymmetry in the subsurface distribution of KREEP layer that is not currently understood and could be addressed by characterizing the incompatible-element signature of rocks from as-yet-unvisited lunar terrains, both by local and regional remote sensing and sample analysis. In another example, isotopic ages of Apollo samples that are understood to be the earliest crustal rocks show significant overlap in ages, inconsistent with a traditional lunar magma ocean view of a primary anorthositic crust later intruded by plutonic rocks. Furthermore, ages of some ferroan anorthosites postdate the age estimates for crystallization of the lunar magma ocean. However, because of the small size and low abundance of radiogenic elements in these rocks, it may be that researchers have not yet sampled a true piece of the pristine lunar crust. More magnesian anorthosite than exists in our sample collection is identified by remote sensing (e.g., in the rings of the Orientale basin) and may represent the primary lunar crust, more tightly bounding researchers' calculations of the magma ocean process and lunar bulk composition.

Scence Goal 3b—Inventory the variety, age, distribution, and origin of lunar rock types.

Researchers base their understanding of the major lunar rock types on Apollo sample knowledge. However, all the Apollo and Luna sample-return sites were within or on the edge of the PKT, and there are no returned samples unequivocally originating from the SPA Basin or FHT (although the lack of KREEP-bearing material in many feldspathic lunar meteorites implies that they come from the Feldspathic Highlands Terrane). Additionally, there are no samples collected from bedrock outcrops from the Moon, so the understanding of their origins is incomplete. The composition or volume of the pristine anorthositic crust of the Moon is not yet known. There is no terrestrial counterpart to KREEP, and the component is elusive as an igneous or plutonic lithology on the Moon. However, this lithology might be the closest thing to an ore on the Moon, where rare elements are concentrated in a specific kind of rock. Some of these rare elements, including U and Th, may be important to future base activities. Understanding how KREEP rocks formed and how they are distributed allows the prediction of where else on the Moon they may be located, even if they are not expressed at the surface. A poorly understood lithology that may be genetically related to KREEP is the magnesian suite of rocks. These rocks are ubiquitous in the Apollo samples and have been thought of as a highland rock type, but it is uncertain whether these rocks are special products of the PKT or represent plutonic activity throughout the lunar crust. Smaller regions contain unique materials that may not be widespread on the lunar surface but can be of great interest to science and in situ resource utilization—for instance, high-Ti basalts and pyroclastic glass deposits.

Science Goal 3c—Determine the composition of the lower crust and bulk Moon.

The concept of the lunar magma ocean depends to a large extent on understanding the composition and structure of the lunar crust and the bulk composition of the Moon. Key to this understanding is knowing the distribution and volume of plagioclase, mafic rocks, and incompatible-element rich rocks (KREEP). Researchers also need to know the extent of variability among these rocks and whether these materials are related to the pristine crust, later intrusive rocks, or differentiates of thick impact-melt sheets. The South Pole-Aitken Basin may have excavated or melted the lower crust of the Moon or may possibly even provide a window to the lunar mantle. Lunar pyroclastic flows may bring deep-seated rocks to the surface. Global, high-resolution multispectral data will help identify the lateral extent of these materials at the surface, improved geophysical data will help constrain vertical distribution of materials, and sample analysis provides direct knowledge of rock types, lithologic associations, chemical compositions, crystallization ages, and depth constraints. In turn, these data will provide important constraints in interpreting seismic and heat flow data.

Science Goal 3d—Quantify the local and regional complexity of the current lunar crust.

Geophysical models of the lunar crust are highly dependent on assumptions about the type and distribution of materials across the crust and at depth. Samples and detailed geologic maps from regions that have exhumed materials from depth will provide further constraints on these models, which in turn can highlight new areas of interest for future exploration. The lunar center of figure is offset from its center of mass owing to a thick crust on the lunar farside or a more dense crust on the lunar nearside. Geophysical measurements such as heat flow and seismic reflection data will be able to pin down the crustal thickness and global remote sensing.

Science Goal 3e—Determine the vertical extent and structure of the megaregolith.

The megaregolith resulted from the bombardment of the early lunar crust by an intense and highly energetic impact flux, including several large bodies that formed the major multiring basins. Large impacts, such as the South Pole-Aitkin Basin impact, could have brought material from as deep as 200 km to the surface and spread it all over the Moon. The early bombardment could have fragmented and mixed crustal materials to depths of kilometers or more. This highly fragmented, partially melted, and subsequently compacted material forms the megaregolith. Although researchers probably have samples of megaregolith in the form of impact breccias, they have not directly sampled this ubiquitous unit of early lunar surface evolution. Detailed studies of ejecta from major basins, along with geophysical probing of the existing megaregolith, will be needed to determine the nature of the megaregolith and its relationship to underlying, unaltered crust.

Understanding when and how the diversity of lunar rocks formed and how they are now distributed allows the prediction of where else on the Moon they may be located, even if they are not expressed at the surface. The gold and silver resources in the American West were not found by chance—rather, prospectors figured out how ore deposits were formed, in what environments, and what their surface expressions were. Then, they were able to find and exploit the mineral resources. In the exploration of the Moon, it will be necessary to use global remote sensing, regional geologic mapping and geophysical measurements, and comprehensive sample analysis as complementary techniques to fully understand the nature of the lunar crust. Sample analysis from well-characterized sites representing new terrains allows high-precision laboratory analyses of petrology, composition, and radiometric ages and the ability to continue experiments for decades. Global compositional information derived from remote sensing extends the knowledge gained from samples across the entire Moon and helps predict areas where more sampling needs to be done. High spectral resolution is needed for mineral identification and abundance estimates, but high spatial resolution is needed to assess the geologic relations between diverse surface materials. By integrating global remote sensing, detailed regional geology, and precise sample studies, a predictive capability is gained that allows smarter choices about where to send future missions.

In summary, understanding the complete range of lunar rock types, their formation, and their lateral and vertical distribution is critical to scientific understanding as well as to lunar exploration.

Concept 4: The lunar poles are special environments that may bear witness to the volatile flux over the latter part of solar system history.

The Moon and Mercury share a microenvironment at their poles that is unique in the solar system. The very small obliquity of these small planets causes topographic depressions near the poles to be permanently shaded from sunlight, allowing them to achieve extremely low temperatures, ranging between 25 K and 80 K;[1] these temperatures are not expected elsewhere in the solar system within the orbit of Neptune and nowhere else on exposed silicate surfaces. The presence of these cold surfaces adjacent to the hot surfaces of the Moon and Mercury may allow cold trapping of volatile material that has impacted or otherwise been introduced to the surfaces of these objects. Any water or other volatile molecule that encounters a cold-trap surface will be permanently trapped with respect to sublimation, depending on the temperature of the trap and the vapor pressure of the species. The trapping process suggests that the lunar (and mercurian) poles may record a history of volatile flux through the inner solar system over the lifetime of the traps. The committee also notes that the polar regions may provide critical resources, such as high concentrations of hydrogen and possibly water, for future exploration. By answering the science issues described below, important information will also be provided for potential in situ resource utilization opportunities. Therefore, studies of the lunar polar environment provide an opportunity for strong synergies between scientific and exploration goals (see Box 3.1). Polar exploration also will provide information for siting, engineering properties, illumination conditions, and other data important for human exploration.

The diversity of potential sources and of transport, trapping, loss, and retention mechanisms suggests that the lunar and mercurian poles are an extremely complex deposit of volatiles. However, this diversity is matched by a near-total lack of data to constrain understanding. In the case of Mercury, radar observations of the poles show the presence of very high radar reflectivity in polar craters for which models indicate very low temperatures. These radar properties are consistent with thick ice, although other volatile material has been suggested. Similar observations of regions of permanent shade on the Moon—about 20 percent of the permanent shade can be observed from Earth—show no similar regions of high radar reflectivity, indicating that the lunar and polar deposits are fundamentally different. This difference demonstrates that very similar physical conditions can give rise to completely different results. Radar observations made by Clementine of the south polar lunar crater Shackleton showed that this crater exhibited radar scattering consistent with a coherent backscatter opposition effect caused by thick ice; recent ground-based observations confirmed the polarization results of Clementine at Shackleton, but also showed that these conditions occur within craters that are illuminated, suggesting that the Clementine results may also be due to roughness.

While the radar results do not positively indicate the presence of ice at the lunar poles, to contribute a measurable radar signal, ice must be thick and relatively pure; thus a lack of radar detection cannot preclude ice. Significant admixture with lunar soil, or the presence of alternating layers of soil and ice, would not necessarily be detected

---

[1]The lunar surface temperature in nonpermanently shaded regions varies night to day from about 100 K to 400 K.

BOX 3.1
Linkages Between Lunar Resource Utilization, Science, and Human Exploration

Minimizing the costs of sustaining an outpost on the Moon requires the use of local resources. Interactions between the field of resource utilization and science lie in the following areas:

1. Exploration for resources based on chemical and mineralogical data employs the same data sets as those obtained for understanding the geochemical evolution of the Moon. The data in hand on the chemistry and mineralogy of the lunar regolith provide inputs to lunar resource process development as well as to the history of the regolith.

2. Extraction of lunar resources requires access to the subsurface (excavation, drilling). These techniques can be used to access lunar environments for scientific studies.

3. Development of products based on lunar resources will utilize special aspects of the lunar environment (particularly vacuum). Scientific understanding of the lunar atmosphere and the behavior of molecules on the lunar surface will be important in understanding the potential for lunar contamination from the resource extraction processes. In turn, the processes themselves may require the preservation of high-vacuum conditions.

4. New materials manufactured in the lunar surface environment (e.g., vacuum-coated surfaces) may lead to new types and applications of materials sciences.

5. Access to relatively inexpensive hydrogen and oxygen (less expensive than bringing them from Earth) may enable more intensive development of experimental biological systems and associated scientific studies.

6. Lunar resource development can enable new capabilities for science. Development of a propellant-chemical energy storage capability (e.g., regenerative fuel cells using lunar O_2 and H_2) may accelerate the potential to establish remote field camps away from the lunar outpost. Excavation technology and the extraction of large volumes of water could enable the emplacement of large subsurface cosmic-ray detectors. Metal and nonmetals extracted from the lunar regolith could enable the development of indigenous power (silicon for photovoltaic devices, metals for energy transmission), enabling an energy-rich environment for science. Some by-products of resource processing (e.g., noble gases) might become useful for the support of science experiments.

Near-term exploration priorities for resource development will focus on understanding the distribution of volatiles in the lunar polar regions. The confirmation of readily extractable hydrogen and oxygen can have a major effect on the exploration strategy itself as well as determine the technologies needed to produce the resources. The key is to understand the lateral and vertical variation of hydrogen (and potentially other useful volatiles) in and outside of permanently shadowed craters, locating the most concentrated and readily useful deposits. Exploration to determine the hydrogen distribution for resource purposes will have many attributes in common with exploration that addresses the scientific questions of where the volatiles came from and how they behave on the lunar surface. Scientific exploration may be more interested in isotopic and small-scale stratigraphic distributions than is resource exploration, but the commonalities of locating and determining concentrations suggest that a common exploration approach will be effective.

Also in the near term, excavation techniques may be developed for resource extraction. Development should consider the potential uses of this technology for scientific studies. For example, some excavation techniques may be used more readily than others for excavating trenches in the regolith for the study of regolith structure. Where large amounts (cubic meters) of regolith are obtained for testing resource-extraction techniques, efforts should be made to separate, catalog, and return representative samples of coarse material unsuitable for resource processing. These can contribute to understanding the geologic evolution of local and distant units on the Moon. In addition, the development of materials transportation systems will require an understanding of the physics of particulate materials in the lunar environment.

Resource utilization will be advanced by having a better understanding of the behavior of volatiles in the lunar environment. It is known that agitating or crushing lunar regolith releases solar wind gases, and the production of volatile materials for utilization will increase the probability of environmental contamination by these volatiles. Experiments should be planned on early missions that explore the migration of volatiles in the lunar environment through active release experiments.

In the longer term, two resource issues are potentially significant. The first is the distribution of hydrogen outside of the polar permanent shadow. Hydrogen contents of mature lunar regolith are in the range of 50 parts per million (ppm) to 150 ppm by weight. In addition, some regolith breccias have larger concentrations of hydrogen. A nonpolar source of hydrogen may become important if hydrogen at the poles proves difficult to access. Development of exploration techniques that can determine hydrogen distributions to ±10 ppm locally will be useful for understanding regolith formation processes and can help identify higher concentrations of hydrogen for resource purposes.

Also, in the long term, as the utilization of lunar resources progresses, it may become necessary to locate richer deposits of less-common elements (e.g., phosphorus). These chemical elements and compounds are likely to be associated with less-common rocks that occur in unusual lunar environments (crater central peaks, excavated lower-crust and mantle materials, and pyroclastics) that will be the focus of exploration activities. In the near term, surveys for unusual rock types, which can be carried out with almost any regolith sample obtainable in very large quantities, will serve both scientific and resource objectives. Orbital compositional data obtained in the near term at high spatial resolution (1 km) will be useful in guiding the exploration for the less-common resources of the Moon.

---

by radar. The Lunar Prospector mission detected a distinct neutron albedo deficit over the poles, indicating the presence of hydrogen at maximum concentrations of 150 parts per million averaged over the large 40 km to 50 km footprint of the measurement. If the hydrogen is situated only in permanent shade, then the concentration of water required to give rise to the measured signal is about 1 percent H_2O by weight. The neutron albedo data also indicate that hydrogen is not located at the immediate surface. As a consequence, the upper several centimeters may lack hydrogen, or may be desiccated if the hydrogen is in the form of water, which is consistent with micrometeorite gardening and loss from the surface of a thicker, but low-concentration, deposit. The neutron albedo and radar results are also consistent with patchy occurrences of ice with much higher concentrations, though still probably not surface deposits.

The current data answer few questions associated with polar volatile deposits. Radar results for Mercury and neutron spectrometer data for the Moon have achieved the zero-order existence proof that cold trapping of hydrogen-bearing volatiles does occur. However, the chemical identity of the polar materials, their sources and evolution, and the explanation for the differences between Mercury and the Moon are entirely unknown. Without new data, the value of the polar deposits for addressing larger issues in planetary science is unknown. The science goals related to lunar polar volatiles that can be addressed with new studies are given below.

Science Goal 4a—Determine the compositional state (elemental, isotopic, mineralogic) and compositional distribution (lateral and depth) of the volatile component in lunar polar regions.

Orbital neutron data from Lunar Prospector indicate that the lunar poles have anomalously high hydrogen abundances compared with other parts of the Moon (see Figure 3.3). However, there is little to no information regarding the chemical or physical form of this hydrogen. Based in part on studies of polar temperatures and the expected surface distribution of the polar volatiles, it has been suggested that the hydrogen is mostly composed of water ice. If the polar deposits are dominantly composed of cometary or asteroidal material, one may additionally expect that other volatile elements or compounds would be present in the lunar cold traps. Alternatively, other studies have suggested that the neutron data can be explained as resulting from solar wind implantation, in which case the hydrogen would not be in the form of water ice. There are a variety of processes that may change in situ

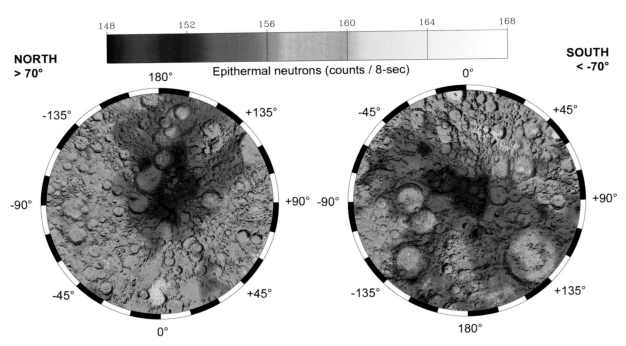

FIGURE 3.3 Polar maps of lunar epithermal neutron counting rates. The dark regions indicate locations of enhanced hydrogen abundances. SOURCE: P. Lucey, R.L. Korotev, J.J. Gillis, L.A. Taylor, D. Lawrence, B.A. Campbell, R. Elphic, B. Feldman, L.L. Hood, D. Hunten, M. Mendillo, S. Noble, J.J. Papike, R.C. Reedy, S. Lawson, T. Prettyman, O. Gasnault, and S. Maurice, Understanding the lunar surface and space-Moon interactions, *Reviews in Mineralogy and Geochemistry* 60:83-220, 2006. Courtesy of the Mineralogical Society of America.

the compositional and physical form of cold trapped volatiles, regardless of how they were initially deposited—for example, forming organics, clathrates, or clay minerals. Finally, except for indications from neutron data that the hydrogen may be buried by tens of centimeters of dry lunar soil, there is very little information regarding the lateral and depth distribution of polar volatiles.

Science Goal 4b—Determine the source(s) for lunar polar volatiles.

Many possible sources for polar volatile material exist. Solar wind directly illuminates the cold traps, and solar wind gas can be trapped, as it is trapped in lunar equatorial soil that experiences temperatures as high as 400 K. Comets, wet asteroids, and interplanetary dust particles can contribute their water and other volatiles through transport and trapping to the poles, or by direct impacts into the permanent shade. Ions are continually stripped off Earth's upper atmosphere, and as the Moon passes through the magnetotail these ions can be trapped at the poles. Unusual and optically very fresh features on the Moon suggest that recent, large-scale, localized outgassing may have provided gases to the poles. Finally, the solar system passes through giant molecular clouds on timescales of tens of millions of years; this extrasolar dust may give up its volatile component to the lunar poles.

Science Goal 4c—Understand the transport, retention, alteration, and loss processes that operate on volatile materials at permanently shaded lunar regions.

Transport of volatiles to the poles occurs as flows through the lunar atmosphere, so transport mechanisms are intimately related to the understanding of the atmosphere. Some measurements exist on volatile flow between the day and night side; no comparable information exists near the poles that could constrain fluxes, net transport, and exchange. While cold trapping will tend to capture volatile material arriving via the lunar atmosphere, several loss mechanisms operate to deplete the cold traps. Lyman alpha ultraviolet radiation scattered off hydrogen throughout the solar system can efficiently erode surface ice, while micrometeorite impact can remobilize vapor molecules, subjecting them to photodissociation in sunlight and solar wind sweeping. The solar wind does impact the permanently shaded regions as it is bent by Earth's magnetotail, subjecting ice in permanent shade to sputtering loss. Paradoxically, the low temperatures may actually promote loss. Surfaces maintaining extremely low temperatures do not allow significant downward migration of volatiles into the regolith, exposing them to loss mechanisms. Intermediate temperatures or regions that experience diurnal temperature variations may be favored for volatile retention.

Meteorite bombardment can sequester volatiles through burial. An optically thick but physically very thin layer of dust protects ice from ultraviolet illumination and sputtering. Volatiles can also be retained if chemically processed in the permanent shade to organics or bound chemically to silicates. Organic production can be stimulated within an ice-bearing silicate regolith by protons produced by cosmic-ray interaction with the lunar surface, or by ultraviolet illumination at the surface, provided temperature cycling is present and sequestration allows this to occur.

The transport of volatile material to the cold trap probably alters the chemical and isotopic composition and relative abundances of volatiles introduced by random impacts on the Moon. Most common cometary volatiles have lifetimes against photodissociation long enough to allow trapping significant quantities at the pole, but the differences in vapor pressure and ionization potential probably cause fractionation en route to the poles.

Finally, studies of transport and alteration processes are also needed for understanding how robotic and human missions to the Moon can affect the pristine lunar polar environment.

Science Goal 4d—Understand the physical properties of the extremely cold (and possibly volatile rich) polar regolith.

The polar regions may provide a natural laboratory for studying physical conditions of planetary or astrophysical importance. Polar silicate grains, perhaps coated with volatiles, are subject to the same radiation encountered by silicate grains in interstellar space. The host of physical changes that occur as the lunar surface is exposed to space weathering can be studied using polar soil that weathers at extremely low temperatures. For example, the extremely low temperatures may inhibit the formation of glass that is ubiquitous elsewhere on the Moon. In addition, if water ice is present and of utilitarian interest, the physical properties of regolith within the permanent shadow will need to be understood, as they will determine the ease or difficulty of excavating regolith for volatile extraction.

Science Goal 4e—Determine what the cold polar regolith reveals about the ancient solar environment.

Polar soil may also address important issues in solar science. Equatorial soil heated to very high temperatures releases deeply implanted, high-energy solar wind gas. This gas has an unusual nitrogen isotopic signature not predicted by solar composition models. However, this signature may be due to thermal diffusion and isotopic fractionation. Polar soil in contrast has never been heated, so diffusion can be eliminated as a process affecting the isotopic composition of trapped solar gas.

Analysis of existing data (provided by Clementine, Lunar Prospector, Earth-based radar) and data from planned orbital missions will contribute significantly to understanding of the lunar polar volatile deposits. For example, NASA's Lunar Reconnaissance Orbiter (LRO) will provide photometry, morphology, topography, and temperature information that will improve knowledge of the polar environment. LRO will also provide information about the hydrogen distribution very near the poles and possible surface frosts if they are present. The Moon Mineralogy Mapper imaging spectrometer on the Indian Space Research Organization's Chandrayaan-1, could potentially detect water or hydroxyl features through measurement of surfaces illuminated by sunlight scattered by nearby topographic highs. Because the richest deposits are most likely to preserve a record of volatile history and evolution with greatest fidelity, it is important that the location and volatile inventory of such deposits are determined by the best methods available, including innovative remote sensing approaches that can characterize volatile abundance to latitudes as low as 70 degrees.

Even so, a positive result by upcoming orbital remote sensing missions is unlikely to answer the most compelling scientific questions; these can only begin to be addressed by in situ measurements. Because abundances of scientific importance are far below orbital detection limits, a negative orbital result does not diminish the potential scientific value of the poles, but it does complicate their further investigation if candidate landing sites are not identified.

First, direct in situ measurements of key regolith volatile characteristics within the cold traps, including abundances and chemical, mineralogical, and isotopic compositions, would provide the first meaningful constraints on models. These measurements should include controls consisting of measurements on polar regolith not in permanent shade. While abundances within the cold traps may be spatially variable, analysis of even trace abundances would be of extreme value, so prior knowledge of the richest deposits is not required for significant progress.

Second, characterization of the lunar atmosphere in proximity to the cold traps, including direction information that can be used to infer flows, is essential to understanding the relationship between the atmosphere and the cold traps. This characterization is required prior to human landings to avoid contamination.

Third, spatial variability of volatile deposits within permanently shaded regions needs to be determined, since there are indications that these regions may be patchy within the shaded regions.

Fourth, for the richest deposits, in situ measurements may need to be made as a function of stratigraphy, since there are indications that these deposits are layered. In addition, the in situ regolith measurements will result in a basic understanding about the physical properties of the regolith in permanently shaded regions for which there is no direct information. Characterization of variations in the ancient solar wind would require sampling several depths in the polar regolith.

The results of these initial measurements would guide further investigations. If it can be shown that lunar polar cold traps preserve volatiles with high fidelity, such that source characteristics can be inferred, or if in contrast the volatiles reveal significant degrees of processing (e.g., to complex organics), subsequent, more-detailed analysis is warranted, likely using cryogenically preserved returned samples.

In summary, a unique opportunity exists to characterize the volatile compounds of polar regions on an airless body and determine their importance for the history of volatiles in the solar system.

Concept 5: Lunar volcanism provides a window into the thermal and compositional evolution of the Moon.

As discussed in Concept 3, understanding of lunar crustal evolution has been tied for many years to somewhat simplified models of the lunar magma ocean, which explained global trends in crustal composition but left many questions unanswered at regional and local scales. The cause of the nearside-farside asymmetry in lunar volcanic activity, for example, remains an unresolved problem. More-complex models of the magma ocean have the potential to address such questions, but they require compositional, temporal, and geophysical constraints to be effective.

Currently, the connections between composition, location, and age of volcanic activities are somewhat limited. On the one hand, the lunar sample collection has yielded detailed composition and age data for a subset of volcanic rocks, but the geologic context needed to interpret them is often lacking. Global remote sensing data, on the other hand, reveals many volcanic rock compositions that do not appear in the sample collection. A return to the Moon offers the opportunity to bridge the gap between the sample collection and the wealth of remote sensing data and to address specifically a number of key goals concerning the understanding of lunar volcanism and the evolution of the Moon.

Science Goal 5a—Determine the origin and variability of lunar basalts.

Approximately 17 percent of the lunar surface is covered by mare basalts, with the majority having erupted on the lunar nearside. Although they are volumetrically a small part of the total lunar crust, the mare basalts provide critical constraints on the differentiation and thermal history of the Moon. Although many different types of mare basalts are represented in the Apollo and Luna collections and among lunar meteorites, key basalt units identified from orbit remain unsampled, and consequently their detailed chemistry and absolute ages remain unknown.

Laboratory analysis of well-chosen samples across a range of maria will address these questions. In particular, samples of basalts that erupted on the lunar farside will help elucidate at what depths melting occurred, when eruptions took place, and whether the composition of the mantle is uniform from the nearside to the farside. A range of subsurface sounding methods will permit determination of the thickness and structure of individual benchmark basalt flows. Investigating the thermal state and history of the interior (e.g., through careful measurements of the interior heat flow) will establish the thermal constraints on magma production through time and help answer fundamental questions about melt generation, segregation, and transport.

Science Goal 5b—Determine the age of the youngest and oldest mare basalts.

Recent crater counts suggest that some of the mare basalts in Oceanus Procellarum might be as young as 1.2 Ga—an age unrepresented anywhere in the sample collection and an important calibration point for understanding of both volcanism and the impact cratering flux (see Concept 1). At the other end of the time line, the oldest mare basalts and their link to "pre-mare volcanism" are not well understood. For example, some fragments of high-Al and the KREEP basalts found in the sample collection are older than the oldest known mare samples, but it is not clear if they are distinct volcanic processes or if they are part of a continuum that evolved into mare volcanism.

Samples of the youngest and oldest basalts will help to address these questions and constrain how basaltic processes have evolved over time.

Science Goal 5c—Determine the compositional range and extent of lunar pyroclastic deposits.

Pyroclastic volcanism offers the most direct sampling of the lunar mantle, as well as immense resource potential with respect to materials such as oxygen, iron, and titanium. Within the existing sample collection, the range of volcanic glass compositions is large, and it is likely that this range will expand even further as new deposits are sampled and their composition and age are assessed. Because of the role of pyroclastic deposits in approximating primary magmas, more examples of these deposits will provide information on the depth of the magma ocean, the character of the lunar mantle, and of course the nature of the mare basalt source regions.

Science Goal 5d—Determine the flux of lunar volcanism and its evolution through space and time.

Neither the magma production rate through time nor the chemical evolution of these magmas and the thermal evolution of the Moon overall are known. Because of the potential importance of high-Ti basalts and pyroclastic deposits as potential lunar resources, increasing the understanding of lunar volcanism in space and time is important from both a scientific and an exploration perspective. Ultimately, this goal includes and expands on those above, with each answered question contributing to the overall understanding of lunar volcanic processes and their products.

Planned or potential future orbital systems offer some improvement in our knowledge, through elemental and mineralogical mapping of volcanic materials, for example, and crater counts from higher-resolution data to pin down relative ages. But detailed modeling is needed to address the connections between volcanic source regions and surface materials, and such models require better geochemical constraints than are currently available. These constraints can be addressed through a range of landed activities. New mare and pyroclastic samples, selected specifically from benchmark deposits and returned to Earth for detailed analyses, would offer dramatic improvement to current petrologic models. In situ analyses by rovers would further expand the range of samples and add information about their geologic context. With the advent of human fieldwork, core samples through whole sequences of lava flows, or subsurface sounding to determine flow volumes, will all offer critical information needed to build a comprehensive picture of lunar volcanic evolution.

Thus, new samples and a variety of in situ measurements will provide a clear view of the overall history of lunar volcanism and its relation to the Moon's thermal and compositional evolution.

Concept 6: The Moon is an accessible laboratory for studying the impact process on planetary scales.

Impact cratering is a fundamental process that affects all planetary bodies. Understanding of cratering mechanics is heavily biased by observations of craters on Earth and in Earth-based laboratories. While this understanding has been scaled as much as possible for lunar gravity, there are many untested hypotheses about lunar cratering, including the detailed structure and rim diameter of multi-ring impact basins, the effects of target composition on crater morphology, the amount of central uplift within craters, the existence and extent of impact-melt sheet differentiation, the mixing of local and ejecta material, and scaling laws for oblique impacts. In this context, the Moon provides unique information because it allows the study of cratering processes over several orders of magnitudes, from micrometeorite impacts on glassy lunar samples, to relatively recent rayed craters (Figure 3.4), to the largest basin in the solar system, the South Pole-Aitken Basin (e.g., see Figure 3.1). The large number of lunar impact craters over a wide range in diameters provides the basis of statistically sound investigations, such as, for example, depth/diameter ratios, which in turn have implications for the possible layering and strength of the lunar crust but can also be extrapolated to other planetary bodies. Thus, the Moon is a valuable, easily accessible, and unique testbed for studying impact processes throughout the solar system.

The National Research Council's decadal survey report *New Frontiers in the Solar System: An Integrated Exploration Strategy* asked, "How do the processes that shape the contemporary character of planetary bodies operate and interact?"[2] Cratering is one of several such processes, affecting the lunar surface, the crust, and possibly even the mantle; each advance in understanding of cratering mechanics moves researchers closer to answering that key scientific question. Current hypotheses and assumptions about cratering processes underpin many of the hypotheses about the composition and evolution of the lunar crust, and thus the rest of the solar system. For example, answering the question of whether or not impact-melt sheets can differentiate will either open or close a door on the range of potential origins of igneous rocks found on the Moon. Some cratering hypotheses are rarely questioned and have become sufficiently accepted that they are now "rules of thumb." An example of this involves the amount of central uplift within craters (e.g., see Figure 3.5). Models of crustal structure and character have been derived from data on the composition of central peaks of lunar craters; if these peaks did not originate from the depths currently assumed, then these models might require re-evaluation. Thus, testing and validating both the wildest and the most accepted hypotheses about the cratering process are critical to being able to correctly interpret lunar geology, and ultimately that of the solar system. To achieve this, sample return from craters and basins, including a vertical sample of a basin melt sheet, is needed. In addition, the walls, rim, and central peaks of compositionally diverse major complex craters need to be mapped in geologic detail, beginning with orbital measurements and followed by selected field studies of at least one crater. The structure of large multi-ring basins needs to be mapped through drilling programs or geophysical measurements.

Science Goal 6a—Characterize the existence and extent of melt sheet differentiation.

Within very short periods of time, impacts transfer enormous amounts of kinetic energy into the target, resulting in shock metamorphism. Above pressures of about 40 gigapascal to 100 gigapascal, whole-rock melting begins, producing so-called impact melts. These impact melts make up about 30 to 50 percent of our sample collection and are extremely important, for example, for dating large lunar impact basins. However, from the terrestrial Sudbury Igneous Complex, it is conceivable that the melt sheet cooled slowly enough to allow differentiation. It is currently unknown if the Sudbury example is a valid analog to large lunar impact basins and whether the lunar melt sheets also underwent significant amounts of differentiation.

Science Goal 6b—Determine the structure of multi-ring impact basins.

Being more than 400 km in diameter, multi-ring basins are the largest impact structures on the Moon and the planets. The definition of diameters for the various basins varies, depending on the exact criteria and data sets

[2]National Research Council, *New Frontiers in the Solar System: An Integrated Exploration Strategy*, The National Academies Press, Washington, D.C., 2003.

SCIENCE CONCEPTS AND GOALS

FIGURE 3.4 Copernicus crater (near the horizon, above), 95 km in diameter, and its ray system as seen from orbit during Apollo 17. Lunar impact craters provide direct information about the excavation, transport, and deposition of materials by a major impact event. SOURCE: NASA Apollo 17 AS17-M-2444.

examined. Recently it has been pointed out that the definition of crater diameters of multi-ring basins in the older literature is problematic and should no longer be used. These papers argue for smaller diameters on the basis of geophysical modeling of gravity-field anomalies associated with these basins. However, the definition of the final "rim-to-rim" diameter of a multi-ring basin is difficult and remains a matter of debate. Large terrestrial complex impact structures are frequently eroded to varying degrees. There are also impacts that have been tectonically modified or buried by post-impact sediments (e.g., Chicxulub, Chesapeake). Contrary to the situation with lunar craters, these craters can only be investigated through drilling programs or geophysical techniques, so their detailed structure is not very well known. For example, in the case of Chicxulub, seismic reflection data are available for a faulted rim area and a topographic peak ring, but a loss of coherent seismic reflections does not allow studying the structural details close to the center.

FIGURE 3.5 The central peaks of Copernicus crater. The blocky mountains in the center of the crater are a few hundred meters in height and contain deep-seated material brought to the surface during rebound from the impact. SOURCE: NASA Lunar Orbiter 2.

Science Goal 6c—Quantify the effects of planetary characteristics (composition, density, impact velocities) on crater formation and morphology.

It is known from Earth that target effects have influence, for example, on the transition diameters between simple and complex craters. Terrestrial impact carters are subject to erosion. While this may be detrimental to establishing morphometric relations, it allows studying terrestrial impact structures at different erosional levels. Together with field observations and drilling data, such studies revealed that central peaks of complex craters are the product of uplift of deeper target lithologies. For terrestrial craters with diameters between 4 km and 250 km, the amount of uplift can be expressed by a simple power law. Attempts to relate the terrestrial data to lunar craters yielded a similar power law. However, this power law only holds for lunar craters that are on the order of 17 km to 136 km, and there are caveats and ambiguities, the resolution of which awaits better data. On Earth, craters in sedimentary target rocks, such as the Ries crater, usually do not show high central peaks. In contrast, craters of similar diameter in crystalline target rocks (e.g., Boltysh, Ukraine) have prominent central peaks. It is assumed that the target properties control the morphological expression of the uplift. Variations in gravity from one planetary body to another also influence the crater morphology. Bodies with lower gravity usually show deeper impacts than those of bodies with higher gravity. All natural impacts are to some extent oblique and have been investigated through computer codes and laboratory experiments. However, verification of computer codes with laboratory experiments remains an open issue. Problems in comparing numerical model with impact experiments stem from the higher impact velocities and larger impact scales of planetary impacts that can not be reproduced in laboratory experiments.

Science Goal 6d—Measure the extent of lateral and vertical mixing of local and ejecta material.

Deposition of ejecta is an important factor in the mixing of lunar surface materials. Because this process is very complex on the scale of individual samples, there is only limited consensus on the nature and absolute extent of such mixing. For the Ries crater, it has been demonstrated that local reworked material increases beyond 1 crater radius and comprises 70 to 90 percent of the total clast population of the breccia deposits at 2 to 3 crater radii. Continuing improvements in remote sensing might yield observational data to calibrate the various estimates of mixing ratios as functions of distance.

In summary, implementation of the Vision for Space Exploration presents an opportunity to characterize the cratering processes on a scale that is particularly relevant to understanding the effects of impact craters on planets.

Concept 7: The Moon is a natural laboratory for regolith processes and weathering on anhydrous airless bodies.

Regoliths, exemplified by the lunar regolith, form on airless bodies of sufficient size to retain a significant fraction of the ejecta from impact events. The regolith has accumulated representative rocks from both local and distant sources since the most recent resurfacing event (e.g., the deposition of lavas or a substantial impact debris layer). It also contains modification and alteration products induced by meteoroid and micrometeoroid impacts, and modifications due to the implantation of solar and interstellar charged particles, radiation damage, spallation, exposure to ultraviolet radiation, and so on. A description of the formation of agglutinates to describe the complexity of regolith processes is shown in Figure 3.6. Knowledge of the processes that create and modify the lunar regolith is essential to understanding the compositional and structural attributes of other airless planet and asteroid regoliths.

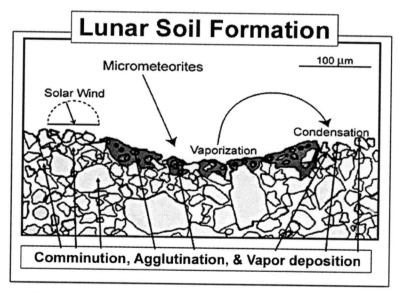

FIGURE 3.6 Micrometeorites impact the lunar soil, some with enough energy to melt the silicate minerals. This melt splashes over grains, quenches to glass, and forms agglutinates. Some melt reaches even higher temperatures and partially vaporizes, only to condense on the surfaces of other grains. SOURCE: Courtesy of Lawrence A. Taylor, University of Tennessee.

Science Goal 7a—Search for and characterize ancient regolith.

Because the regolith collects the products of the interaction of impactors and radiation with the surface, the composition of ancient regoliths, protected by overlying layers of volcanic materials, may yield information on the time-history of the Sun and interstellar particle fluxes in the inner solar system. Layers of interspersed volcanic rocks and ancient regolith can be observed or inferred in the maria (for example, by the Apollo 17 radar sounder, albeit at greater depths than relevant here), where the periods between successive volcanic flows were periods of new regolith formation. Such layers can be investigated from several perspectives: (1) they contain a record of solar particle irradiation at specific times that can be dated by age determination of underlying and overlying basalts, (2) they accumulated materials that were being ejected from lunar surface petrologic provinces at specific past times, and (3) they contain a cumulative record of the composition of impactors on the Moon at those times. The probability of finding meteorites from the ancient Earth will be greater in regoliths of older age.

Sampling of these ancient regolith layers can be carried out by drilling through interspersed volcanic rock and regolith or by collecting rocks in the walls of impact craters or along rilles. If the ancient regolith layers have been indurated through the thermal effects of the overlyig lavas, they might also be discovered in the rock fragments that surround impact craters. Samples may be available at many mare locations, but targeted collection would benefit from on-site human field observations to identify and retrieve the desired sample materials. For example, a sampling device (e.g., a rover) could sample the stratigraphic column in an impact crater wall or in a rille, with an astronaut making critical observations of the properties of the layered sequence using handheld sensors.

Science Goal 7b—Determine the physical properties of the regolith at diverse locations of expected human activity.

Most lunar resources will be derived from the regolith. Understanding the mineralogy, volatile concentrations, and physical properties of the regolith and having a better understanding of regolith formation and history will be crucial to exploring for and developing extractive techniques for regolith-based resources. This is particularly true for the polar regions, for which there is no current basis for understanding these properties in detail. The physical properties of the regolith will be essential information for the most effective design of surface structures for use by human explorers. Better determination of the physical properties of the regolith as a function of depth (strength, cohesion, and so on) will be important in the design of processes and technology to excavate and transport lunar regolith for purposes of radiation protection and resource extraction.

The Apollo drill cores were able to reach 3 meters (m) into regoliths that are 6 m to 10 m in thickness. The bottom of the regolith, at its contact with underlying bedrock, is unexplored, as are the physical properties, particularly in terms of fragment size and layering, of the earliest regolith. This may have a bearing on recognizing ancient regolith layers.

Drill holes in the regolith have other uses as well. The 3 m drill holes of Apollo were used to emplace a heat flow experiment. The deeper the drill hole, the more precise the thermal data could be for heat flow determinations, with a 10 m hole representing perhaps an optimal choice with respect to attainability and the expected thickness of fine-grained regolith materials. Measurement of the borehole temperature profile and the thermal diffusivity of the regolith in a 10 m hole will enable small variations in the past solar intensity to be detected against the steady gradient expected from a constant insolation. Interpretation of a lunar borehole thermal profile is straightforward, providing a clean test of proposed scenarios of solar variations extending back several hundred years. This information about the Sun, derivable from the Moon's regolith temperature profile, has important implications for interpretation of Earth's currently observed global warming.

Science Goal 7c—Understand regolith modification processes (including space weathering), particularly deposition of volatile materials.

Understanding the "space weathering" processes that affect the regolith, particularly the distribution of materials volatilized by impacts, is essential to the interpretation of spectral data used to map the distribution of rock types on the surface. Because the effects of space weathering depend on both the composition and the exposure history of the surface, additional lunar regolith samples that represent materials of different initial composition and age should be prime candidates for study and return to Earth for detailed study. It may be possible to expose artificial surfaces to the lunar environment to examine the effects of micrometeorite impact and volatile deposition. These data will be valuable for the interpretation of remote sensing data of other places on the Moon as well as for other airless bodies.

Science Goal 7d—Separate and study rare materials in the lunar regolith.

The regolith collects fragments of rocks that come to the Moon as meteorites as well as impact ejecta from across the lunar surface. Owing to the discovery of meteorites from the Moon and Mars on Earth, it is now believed that there has been an exchange of materials between the planets over time. There is no reason to believe that the larger size of Earth or its atmosphere precludes the ejection of materials from Earth, some of which could impact the Moon. Modeling of this phenomenon is a new opportunity for scientists, with the possibility that the models can be verified by the discovery and characterization of terrestrial meteorites on the Moon. The probability of finding such material is small; however, a large quantity of lunar regolith is available and could be sampled if simple, rapid screening techniques are developed. Discovery of pieces of ancient Earth rocks on the Moon could provide a new window into early Earth history.

Likewise, the regolith collects samples of rocks and glasses ejected by impact events all over the Moon, the ejecta from the closest impact event being the most likely to accumulate at a given site. A systematic study of a large amount of regolith, selecting rock fragments of a few millimeters diameter, which are large enough for detailed petrologic and geochemical study, should yield a good sampling of the diversity of lunar surface materials. Techniques to identify rarer or previously unsampled rocks are needed. The correlation of regolith rock types with surface exposures may be possible through detailed spectroscopic analysis in concert with remote sensing data.

In summary, through regolith studies during implementation of the Vision for Space Exploration, multiple opportunities may be addressed that will constrain processes involved in regolith evolution, decipher ancient lunar environments, contribute to understanding the history of the Moon, and provide important information for future human activity on the Moon. It may be possible to deduce recent past solar variability from regolith thermal measurements.

Concept 8: Processes involved with the atmosphere and dust environment of the Moon are accessible for scientific study while the environment remains in a pristine state.

Science Goal 8a—Determine the global density, composition, and time variability of the fragile lunar atmosphere before it is perturbed by further human activity.

The nearest example of a surface boundary exosphere (SBE) is the lunar atmosphere. SBEs are tenuous atmospheres whose exobase is at the planetary surface. Because individual atoms and molecules rarely collide in SBEs, kinetic chemistry is all but nonexistent, but important structural and dynamical processes do occur. SBEs are known to exist on Mercury, Europa, Ganymede, Callisto, and Enceladus; they are expected to exist on many other satellites and perhaps even Kuiper Belt objects. SBEs are the least-studied and least-understood type of atmosphere in the solar system but could provide new insights into surface sputtering, meteoritic vaporization processes, exospheric transport processes, and gas-surface thermal and chemical equilibration. The lunar atmosphere is the only SBE atmosphere in the solar system that is sufficiently accessible that researchers can expect to study it in detail using both lander and orbital techniques.

Apollo Lunar Surface Experiments Package (ALSEP) surface station instruments revealed that the mass of the native lunar atmosphere is on the order of 100 tons (3×10^{30} atoms, equivalent to ~10^{11} cm^3 of terrestrial air at sea level [i.e., a cube of terrestrial air roughly $50 \times 50 \times 50$ cubic meters at standard temperature and pressure]). Yet ALSEP total lunar atmosphere mass measurements failed to identify a census of species that comes anywhere close to the total mass of the lunar atmosphere: in fact, over 90 percent of the molecules in the Moon's atmosphere are currently compositionally unidentified.

As a result of its low mass, the lunar atmosphere is incredibly fragile. A typical lunar surface access module (LSAM) landing will inject some 10 to 20 tons of non-native gas into the atmosphere, severely perturbing it locally for a time that might range from weeks to months. A human outpost might see sufficient traffic and outgassing from landings, lift-offs, and extravehicular activities (EVAs), for example, to completely transform the nature of this pristine environment. *For this reason, the committee recommends a strong early emphasis on studies of the native lunar atmosphere.*

The key scientific questions to address are the following: What is the composition of the lunar atmosphere? How does it vary in time with impacts, diurnal cycles, solar activity, and so on? What are the relative sizes of the sources that create this atmosphere and the sinks (loss processes) that attack it?

Science Goal 8b—Determine the size, charge, and spatial distribution of electrostatically transported dust grains and assess their likely effects on lunar exploration and lunar-based astronomy.

During the Apollo era it was discovered that sunlight was scattered at the lunar terminator giving rise to "horizon glow" and "streamers" above the surface. These phenomena were most likely caused by sunlight scattered by electrically charged dust grains originating from the surface, which is itself electrically charged by the local plasma environment and the photoemission of electrons by solar ultraviolet radiation.

Under certain conditions, the like-charged surface and dust grains act to repel each other, thus transporting the dust grains away from the surface. The limited observations of this phenomenon, together with laboratory and theoretical work, suggest that there are two modes of charged-dust transport: "levitation" and "lofting," both of which are driven by the surface electric field. Micron-scale dust is levitated at about 10 cm, while <0.1 micron dust is lofted to altitudes higher than 100 km. The Apollo 17 Lunar Ejecta and Meteorites surface experiment directly detected the transport of charged lunar dust traveling at up to 1 kilometer per second. The dust impacts were observed to peak around the terminator regions, thus suggesting a relationship with horizon glow.

It is necessary to make targeted in situ measurements of dust-plasma-surface interactions on the Moon in order to fully understand this critical environment. The plasma, electric-field, and optical measurements that are required for characterizing the lunar dust-plasma environment can be made from orbit to give a global-scale view and from the surface to give a local view. To optimize the characterization of this environment, the committee recommends that measurements from orbit and from the surface be coordinated, so that the connection between processes at these scales can be understood. Several landers would be advantageous, since not every point on the lunar surface experiences the same conditions—for example, locations near the poles will be quite different from those nearer the equator.

Astronauts could be used to distribute a network of sensors on the lunar surface. In addition to measuring the natural environment, the instrumentation should also detect the charge on the astronauts and the dust transport caused by their moving around on the surface. These measurements will reveal how astronauts and equipment are coupled to the dust-plasma environment. From the experiences of the Apollo astronauts, it is known that dust will be a significant impediment to surface operations; therefore, it is crucial that a much better understanding of this environment be obtained as early as possible.

Science Goal 8c—Use the time-variable release rate of atmospheric species such as ^{40}Ar and radon to learn more about the inner workings of the lunar interior.

The first detection of individual atmospheric species came from the ALSEP and the scientific instrument module (SIM) instruments in the orbiting Apollo service module bay. Among the species discovered by Apollo missions were ^{40}Ar, Po, Pb, Ra, and Rn, all of which emanate from the lunar interior via outgassing. Through the time variability and spatial location of such species, the lunar atmosphere represents a window into the workings and evolution of the lunar interior, including perhaps fractionization and a molten core. After Apollo, ground-based observers detected the alkali tracer species Na and K whose density ratios were close to the lunar surface ratio, suggesting that part of the atmosphere originates from the vaporization of surface minerals by processes such as solar wind sputtering and micrometeorite impact. Na and K are also present in the SBE atmospheres of Mercury, Io, and other Galilean satellites, thereby strengthening the utility of lunar SBE studies for enhancing knowledge of similar atmospheres across the solar system.

Science Goal 8d—Learn how water vapor and other volatiles are released from the lunar surface and migrate to the poles where they are adsorbed in polar cold traps.

Evidence for volatile species, including H_2O, CO, CO_2, and CH_4, was found sporadically by Apollo sensors, but these detections remain unconfirmed. The detection and study of volatiles are of great scientific interest and have obvious implications related to the trapping of ices that could represent resources to be exploited at the lunar poles. The expected sources of volatiles include comets, the solar wind, and meteoroids. Sinks include photodissociation, Jeans escape, solar wind pickup, and condensation. Once released from the surface by heating, sputtering, or other processes, the volatiles perform ballistic hops in a random walk across the surface. As the terminator is approached, the hops get smaller until the volatiles are adsorbed on the surface in darkness only to be released again at dawn. Modeling shows that there is a net migration toward the poles where the volatiles may be condensed in permanently shadowed depressions or craters. Future measurements should be designed to determine what processes (thermalization, release rate, and velocity) control atmospheric migration and what the efficiency of transport to the poles is.

Early observational studies to address these issues and the concern over human-induced modification of the ancient, native lunar environment should include the following:

• A complete census and time variability of the composition of the lunar atmosphere;
• Determination of the size, charge, and spatial distribution of electrostatically transported dust grains;
• Determination of the time variability of indigenous (e.g., outgassing, sputtering) and exogenous (e.g., meteorite and solar wind) sources (see Figure 3.7);

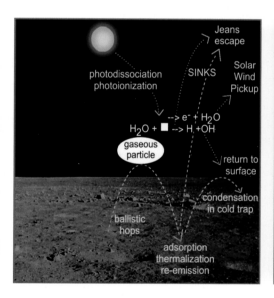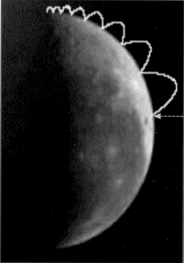

FIGURE 3.7 Lunar volatile transport. SOURCE: D.H. Crider and R.R. Vondrak, The solar wind as a possible source of lunar polar hydrogen deposits, *J. Geophys. Res.* 105:26773, 2000, ©2000 American Geophysical Union. Reproduced by permission of American Geophysical Union.

- Determination of the average rate of volatile transport to the poles, including sunrise/sunset dynamics; and
- Determination of typical loss rates by various processes (e.g., photoionization, surface chemistry, Jeans escape, Michael-Manka mechanism).

Such studies could be completed from early surface networks, fixed or mobile landers, or orbiters, or a combination of any two, with experiments that would include ion-mass spectrometers, optical/ultraviolet spectrometers, and cold cathode gauges. Later, as rocket traffic and human activities perturb the lunar atmosphere from its native state, studies of the environmental effects of human and robotic activity would be highly illuminating, as an "active experiment" in planetary-scale atmospheric modification.

Before extensive human and robotic activity alters the tenuous lunar atmosphere, it is important to understand its composition, transport mechanisms, and escape processes. At the same time, the lunar dust environment must be well characterized so that effective human exploration and astronomical observations can be planned.

4

Implementation

The science concepts discussed in Chapter 3 encompass broad and multicomponent science issues and most of the concepts require a diversity of approaches along with integrated analysis. Table 4.1 tabulates implementation options for each of the eight concepts. Each column of the table provides examples of different types of endeavors that can be achieved for a specific choice of implementation options (information extraction; orbital measurements; sample return; landed experiments, instruments, and rovers; and human fieldwork, or human-tended surface experiments).

Science objectives requiring a permanent human presence or large structures are more aligned with the long-term Vision for Space Exploration (VSE) activities on the Moon and are beyond the scope of this report. Such science opportunities and their requirements should nevertheless be continually evaluated as the VSE is implemented. As discussed in Chapter 7, NASA's VSE program builds on considerable strength if science is fully integrated in operations planning.

Exploration of the Moon is no longer in the reconnaissance phase. As a result of Apollo, we understand that the Moon is a differentiated planetary body, it contains few volatiles, its rocks are old, and its history is closely tied to that of Earth. As discussed throughout the preceding chapters, we now pose far more sophisticated questions about how planets work. Answers to such questions, however, are not simple, nor are they necessarily easy to obtain. As has been the case for the exploration of Mars over the past decade, multiple avenues of implementation for lunar science will be the hallmark of a visionary and progressive science program.

IMPLEMENTATION OPTIONS

Guidelines on how the lunar science concepts might be addressed with different possible elements of the VSE are provided in Table 4.1. Since four sophisticated orbital remote sensing missions are scheduled to return data before the end of 2008 (SELENE [Selenological and Engineering Explorer], Chang'e, Chandrayaan-1, LRO [Lunar Reconnaissance Orbiter]—see below), column (a) of Table 4.1 identifies information and knowledge that can and should be harvested from this rich bounty of orbital data. Assuming that all four of these missions and their host of modern sensors are successful, plans for *information extraction* must be made in order to benefit from the deluge of raw data returned.

Column (b) identifies science return that would result from additional *orbital measurements* beyond those planned in the four missions already under way. It should be noted that if an instrument or mission fails to return

48

TABLE 4.1 Implementation Options for Principal Science Concepts

Science Concepts — The science goals for each concept are discussed in detail in the text (see Chapter 3).	Implementation				
	(a) Information Extraction — An enabling new framework for lunar exploration will be provided by data from SMART-1, SELENE, Chang'e, Chandrayaan-1, and LRO. The assumption is that all missions and key instruments will be successful.	(b) Orbital Measurements — Orbital measurements are not included in the complement of missions planned for launch by 2008. The assumption is that the four missions planned will return appropriate data as planned; if not, new measurements that provide similar high-priority compositional and geophysical data need to be acquired.	(c) Sample Return — The types of returned samples and of science analyses required are identified.	(d) Landed Experiments, Instruments, and Rovers — These include science measurements for/from landed sites; category also encompasses penetrators/impactors.	(e) Human Fieldwork — Science areas that specifically benefit from human capabilities are identified.
1. The bombardment history of the inner solar system is uniquely revealed on the Moon.	Crater counts of benchmark terrain using high-resolution images.	Targeted higher-resolution images of specific terrains.	Sample return from the impact-melt sheet of SPA, from young basalt flows, and from benchmark craters (e.g., Copernicus and Tycho).	Development of in situ instrumentation for dating.	Field observations provide critical geologic context; human interaction improves chances of obtaining best/most appropriate samples.
2. The structure and composition of the lunar interior provide fundamental information on the evolution of a differentiated planetary body.	Farside gravity. High-quality topographic information. Possible information on heat flow and magnetic sounding results.	Relay orbiter for farside stations (e.g., relay of seismic data).	Samples from the interior are important constraints on lunar geochemistry and geophysics (e.g., remanent magnetism).	Simultaneous, globally distributed seismic and heat flow network. Expanded retroreflector network.	Although some landed experiments can be emplaced autonomously, it is assumed that more capable sensors are possible with human guidance/assistance.
3. Key planetary processes are manifested in the diversity of lunar crustal rocks.	Detailed global elemental and mineralogical information in a spatial context. Search for and documentation of a diversity of rock types using returned samples and lunar meteorites. Perform high-resolution mapping of lunar crustal magnetic fields.	Higher-spatial-resolution compositional data are desirable from priority targets. Relay orbiter for farside stations (e.g., relay of seismic data). Magnetic survey from 10 km orbit.	Return samples from priority targets. Every return mission should include a bulk soil and a sieved sample with geologic documentation.	Strategic site selection. Conduct in situ analyses and mineralogical and elemental characterization of the rocks and provide a thorough description of the geologic context. Determine the vertical structure using an active regional seismic network.	Field observations provide critical geologic context; human interaction improves chances of obtaining best/most appropriate samples.

Science Concept					
4. The lunar poles are special environments that may bear witness to the volatile flux over the latter part of solar system history.	Primary understanding of polar environment (photometry, morphology, topography, temperature, and distribution of volatiles).	High-spatial-resolution distribution of volatiles on and in the regolith poleward of 70 degrees.	Cryogenically preserved sample return to determine the complexity of the polar deposits.	Understand physical properties of polar regolith. Determine the localized character and lateral and vertical distribution of polar deposits. Measure chemical and isotopic composition and physical and mineralogical characteristics.	Human-assisted robotic exploration of regolith.
5. Lunar volcanism provides a window into the thermal and compositional evolution of the Moon.	Detailed global elemental and mineralogical information in a spatial context. Improved age-dating for basalts through crater counting.	Stratigraphy of specific basalt flows (subsurface sounding). High-spatial-resolution compositional data desirable.	Sample the youngest and oldest basalt flows. Need samples from unsampled benchmark lava flows and pyroclastic deposits.	Strategic site selection. Conduct in situ analyses and mineralogical and elemental characterization of the rocks and provide a thorough description of the geologic context.	Strategic site selection, core drilling, and active subsurface sounding to determine layering and volume. Sample a complete sequence of flows to determine the evolution of basalt composition.
6. The Moon is an accessible laboratory for studying the impact process on planetary scales.	Detailed geologic mapping of compositionally diverse craters and basins.	Evaluation of upper-surface stratigraphy (sounding). Determination of the shape of craters and the distribution of ejecta.	Sample returns from benchmark craters and basins.	In situ compositional and structural analyses of craters and basins (via traverses).	Core samples from impact-melt sheets. Traverses across ejecta blankets.
7. The Moon is a natural laboratory for regolith processes and weathering on anhydrous airless bodies.	Maps of regolith maturity and derivation of the temporal progression of space weathering. Identification of regions that contain ancient regolith.	Evaluation of upper-surface stratigraphy (sounding).	Regolith from unsampled terrain of diverse composition and age. Understand the evolution of the regolith. Sample old regolith where it is stratigraphically preserved.	Characterization of returned sample environment.	Obtain paleoregolith samples (exposed in selected outcrops or through deep drilling).
8. Processes involved with the atmosphere and dust environment of the Moon are accessible for scientific study while the environment remains in a pristine state.	Characterize surface electric field; dust grain size, charge, and spatial distribution, and effects of human activity on dust environment.	Variation in mass with time and compositional inventory ("with time" refers to the lunar diurnal and Earth-orbital/solar cycles).	Not applicable. Sample return not currently feasible.	Variation of mass with time and identification of dominant species. Environmental monitoring near human activity. Measure electric field and dust environment.	Not applicable. Human presence will alter atmospheric characteristics.

NOTE: Acronyms are defined in Appendix B.

the quality of data anticipated, reflight of comparable instruments is required in order to acquire data that will address items in column (a).

Column (c) provides examples of the type of materials that would specifically benefit the different science concepts through *sample return* and analyses in Earth-based laboratories. Analytical capabilities improve with each technology advancement, continually expanding the value of the sample-return investment.

Column (d) highlights an array of *landed experiments, instruments, and rovers* that will contribute substantially to exploration of the Moon in specific terrains. In situ activities are fundamental to detailed scientific understanding.

Initial human activities on the Moon fall within the timeframe of this report, and column (e) provides examples of *human fieldwork* to be undertaken for each science concept. These are activities that specifically benefit from the abilities of humans present to carry out integrated or challenging tasks. Well-designed human-robotic partnership will be central to the success of the activities.

The actual implementation of individual options within NASA's VSE requires an integrated partnership between universities, NASA and government centers, industry, and the private sector. If the United States wishes to take a leadership role in this activity, then sustained commitment must be made to involve each of these partners in the effort and to maintain and build on strength and experience developed in the U.S. science and engineering communities. In addition, it is clear that an expanding group of space-faring countries will continue to play a central role in exploration of the Moon, and ultimately in how lunar resources are used in human society. Developing the appropriate balance and interaction between U.S. participants and foreign partners/collaborators will be a challenge as well as an opportunity.

INTERNATIONAL CONTEXT

The lunar exploration activities of the recent past and the near future are pervasively international in scope. The European Space Agency (ESA) launched Small Missions for Advanced Research in Technology (SMART)-1 to the Moon in September 2003 on a technology-demonstration mission to validate solar-electric propulsion systems. After a long journey, SMART-1 entered orbit around the Moon and began limited studies of the lunar surface with a suite of small, innovative instruments. SMART-1 scheduled a successful end-of-mission impact on the lunar nearside along with coordinated observations during the fall of 2006. A new image mosaic from the farside of the Moon obtained by SMART-1 is shown in Figure 4.1.

The Japanese Aerospace Exploration Agency has planned two missions for near-term implementation, Lunar A and SELENE. Lunar A is designed to study the lunar interior using seismometers and heat flow probes deployed by penetrators, but technical difficulties during testing put the mission on hold and it was later cancelled. SELENE, however, is a mature orbiter prepared for launch in 2007 for a 1-year nominal mission. The goals of SELENE are to study lunar origin and evolution and to develop technology for future lunar exploration. It carries an array of modern remote sensing instruments for the global assessment of surface morphology and composition. SELENE also carries two subsatellites that will enable the gravitational field of the farside to be measured accurately.

The Chinese National Space Administration formally announced its Chang'e lunar program in March 2003. Chang'e 1, a lunar orbiter with a broad complement of modern instruments, is prepared for launch in 2007. Chang'e 1 carries several remote sensing instruments to study surface topography and composition as well as the particle environment near the Moon. In addition, Chang'e 1 carries a four-wavelength microwave sounder to probe the regolith structure. Future elements being planned for the Chang'e program include a lander/rover and a sample-return mission as precursors to human exploration.

The Indian Space Research Organization will launch its Chandrayaan-1 spacecraft in 2008 on a 2-year orbital mission to perform simultaneous composition and terrain mapping using high-resolution remote sensing observations at visible, near-infrared, x-ray, and low-energy gamma-ray wavelengths. This spacecraft will carry two sophisticated instruments from the United States to characterize and map mineralogy using near-infrared spectroscopy and to map the shadowed polar regions by radar. It will also carry three ESA instruments, two of which were prototypes on SMART-1. In addition to the remote sensing experiments, the Chandrayaan-1 spacecraft will also carry an instrumented probe that will be released and targeted for a hard surface landing.

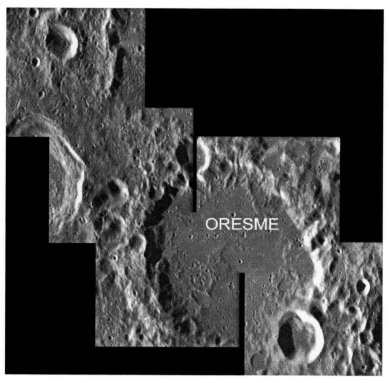

FIGURE 4.1 Image mosaic across the northwest part of South Pole-Aitken Basin obtained by the Advanced Moon Micro-Imager Experiment camera onboard the SMART-1 spacecraft. The subdued 76 km crater Oresme is of Nectarian age. The small, sharp, kilometer-scale features to the northwest that are oriented in several directions are of unknown origin. SOURCE: Courtesy of the European Space Agency/SMART-1 and the Space Exploration Institute.

NASA's Lunar Reconnaissance Orbiter is scheduled for launch in the fall of 2008. LRO's goals are to improve the lunar geodetic network,[1] evaluate the polar areas, and study the lunar radiation environment. A secondary payload, Lunar Crater Observation and Sensing Satellite (LCROSS), launched with LRO, will result in an impact into a polar region target with coordinated analysis. After the first year of measurements, the LRO instruments will be operated to maximize the science return. NASA's plans for the next step of robotic exploration are currently not specified.

The group of four missions described above, having highly sophisticated sensors, will produce an unprecedented array of exceptionally valuable data for the Moon. There are several unique instruments on each spacecraft that give each mission its own flavor and scientific emphasis. There are also a number of similar instruments on different spacecraft that provide an excellent opportunity for cross calibration and validation between missions. All of these international participants in lunar exploration have expressed their intention of publicly releasing data returned (typically 1 year later) in a compatible format that will allow fruitful comparisons and the planning of international lunar exploration.

As we embark on this new era of detailed lunar exploration, several other nations have expressed a serious intent to participate with additional orbital spacecraft sent to the Moon in the near future: Russia, Germany, Italy, and perhaps Great Britain and Ukraine. After the initial orbital missions of Japan, China, and India, the intended next steps by these nations have been publicly stated to be landed in situ experiments or sample return.

---

[1]Establishing a precision geodetic coordinate system is essential for the cartographic needs of both exploration and science. With the tremendous amounts of data expected to begin arriving with LRO (over 350 terabytes), setting a uniform standard for lunar data sets is essential and urgent, given the LRO launch date.

5

Prioritized Lunar Science Concepts, Goals, and Recommendations

According to the statement of task for the Committee on the Scientific Context for Exploration of the Moon (see Appendix A):

> The current study is intended to meet the near-term needs for science guidance for the lunar component of the Vision for Space Exploration. . . . [T]he *primary goals* of the study are to:
>
> 1. Identify a common set of prioritized basic science goals that could be addressed in the near-term via the LPRP[1] program of orbital and landed robotic lunar missions (2008-2018) and in the early phase of human lunar exploration (nominally beginning in 2018); and
> 2. To the extent possible, suggest whether individual goals are most amenable to orbital measurements, in situ analysis or instrumentation, field observation or terrestrial analysis via documented sample return.

The committee based its guidelines for setting science priorities on those outlined in the National Research Council's decadal survey *New Frontiers in the Solar System: An Integrated Exploration Strategy:*

a. Scientific merit (most important). This guideline includes the degree to which an activity will test or alter an existing paradigm or prevalent hypothesis, the question of whether or not the new knowledge will have a pivotal effect on future science endeavors, and whether the new knowledge is likely to expand the factual basis of our understanding significantly.

b. Opportunity and realism for achieving a goal. This guideline addresses whether an activity is likely to produce the desired result in the time frame specified and whether the opportunity readily exists to address the goal.

c. Technological readiness. This guideline concerns whether the technology necessary to carry out the activity is available or anticipated in the time frame specified.

New Frontiers in the Solar System discusses and prioritizes individual mission concepts for the exploration of the solar system in different cost categories. In contrast, this committee does not have comparable guidelines or bounds on how science might be implemented on and for the Moon. It has no information on what resources might

---

[1]The Lunar Precursor and Robotic Program (LPRP) was how robotic missions were identified in the NASA letter that requested this study. The LPRP terminology is no longer in use.

be available or when. The near-term robotic architecture for implementation of the Vision for Space Exploration (VSE) remains blank after the planned launch in the fall of 2008 of the Lunar Reconnaissance Orbiter (LRO).

In identifying prioritized basic science goals as requested in its statement of task, the committee structured its prioritization along three lines: (1) prioritization of the science concepts presented in Chapter 3, (2) prioritization of science goals identified in Chapter 3, and (3) specific integrated high-priority findings and recommendations. The prioritization is based on the consensus of the members of the committee after detailed discussion in each of these three areas. Although the rationales for the prioritization of the items in these three areas are linked throughout the discussion of this report, the implementation requirements are described in broad terms for the science concepts and in more specific terms for the eleven highest priority science goals. The committee reiterates that its priorities and recommendations relate to the near-term implementation of the Vision for Space Exploration, which includes the robotic precursors and initial human excursions on the Moon. The committee sets out a candidate lunar research strategy for the near term in Box 5.1. Planning for and implementing longer-term scientific activities on the Moon are beyond the scope of this study.

BOX 5.1
Candidate Lunar Research Strategy for the Near Term

The discussions and deliberations of the Committee on the Scientific Context for Exploration of the Moon can be consolidated into a near-term candidate lunar science strategy, which would fit into the time interval 2010-2022, the period after the Lunar Reconnaissance Orbiter (LRO) mission but with some overlap of the early phases of the projected Lunar South Pole Outpost described in the preliminary (2006) NASA Lunar Architecture. The committee provides here a set of preliminary concepts for activities that could be implemented by NASA. The following, which are the five highest integrated science implementation priorities that emerged from committee discussions, could be addressed:

1. *Utilize information from Apollo and post-Apollo missions or upcoming lunar science missions (U.S., as well as international) to the fullest extent.* This is a low-cost/high-return element of the lunar science program. Should there be a major series of failures among the missions now projected (see Chapter 4 of this report), fill the critical information gaps with a back-up lunar orbiter mission.

2. *Conduct a robotic landed mission to explore the lunar polar environment.* Determine the nature and source of volatiles within shadowed craters near one of the lunar poles, assess lunar polar atmospheric properties, and emplace a geophysical package that could include seismometer and heat flow experiments.

3. *Emplace a geophysical network to include, at a minimum, seismic and heat flow experiments, environmental sensors, and new laser ranging retroreflectors.* Such a program should be coordinated with those of other countries that are likely to include lunar landed missions in their space exploration strategies. The minimum number of landed sites should be four, more or less equidistantly placed, including at least one farside site (no retroflector required).

4. *Conduct two or more robotic sample-return missions:*
   —The unique nature of the South Pole-Aitken (SPA) Basin makes this area an appropriate first target for a sample-return mission to explore central locations of the SPA Basin (a two-lander scenario was studied as a New Frontiers mission in 2006). Proper placement of these missions could assess quite old mare basalt units and melt rocks from basins that formed within the SPA Basin subsequent to the SPA Basin event.
   —Use technology developed for the SPA Basin sample-return mission to collect samples from the youngest volcanic terrain on the Moon. Many sites that are not likely to be visited soon by astronauts could be accessed with this capability, including missions that could be carried out after humans land on the Moon.

5. *Conduct detailed exploration of the lunar crust as exposed in or near a South Polar human lunar outpost.* The South Pole is on the periphery of the SPA Basin, so correlation between these two areas of sample studies would be valuable. The human mission should include appropriate field investigations, geophysics, and atmospheric investigations and could follow up on the results of an earlier robotic mission, noted above, to a shadowed crater.

TABLE 5.1 Implementation Options for Highest-Priority Science Goals

| Science Goals | Implementation | | | | |
	(a) Information Extraction — An enabling new framework for lunar exploration will be provided by data from SMART-1, SELENE, Chang'e, Chandrayaan-1, and LRO. The assumption is that all missions and key instruments will be successful.	(b) Orbital Measurements — Orbital measurements are not included in the complement of missions planned for launch by 2008. The assumption is that the four missions planned will return appropriate data as planned; if not, new measurements that provide similar high-priority compositional and geophysical data need to be acquired.	(c) Sample Return — The types of returned samples and of science analyses required are identified.	(d) Landed Experiments, Instruments, and Rovers — These include science measurements for/from landed sites; category also encompasses penetrators/impactors.	(e) Human Fieldwork — Science areas that specifically benefit from human capabilities are identified.
1a. Test the cataclysm hypothesis by determining the spacing in time of the lunar basins.	Continue geochronology of impact-melt rocks in the Apollo and meteorite collections. Use remote sensing to help determine the regional geologic context of returned samples.	Higher-resolution images to provide targeted crater counts on selected ejecta facies.	Sample return from the SPA Basin melt sheet and from floors or ejecta of basins within the SPA Basin for detailed geochemical and isotopic analyses.	Develop instruments for precise, in situ geochronology. Use landed geochemical instrumentation to identify best samples for return.	Identify and acquire samples of impact-melt rocks in the Nectaris basin.
1b. Anchor the early Earth-Moon impact flux curve by determining the age of the oldest lunar basin (South Pole-Aitken Basin).	Search for SPA Basin materials in existing collection. Continue study of the ancient terrestrial crater record through fieldwork and zircon evidence.	Higher-resolution images to provide targeted crater counts on ejecta of basins within the SPA Basin to bound a limit on the SPA Basin age.	Sample return from the SPA Basin melt sheet and from floors or ejecta of basins within the SPA Basin for detailed geochemical and isotopic analyses.	Develop instruments for precise, in situ geochronology. Use landed geochemical instrumentation to identify best samples for return.	Human geologic fieldwork increases chances of recognizing the best samples.
1c. Establish a precise absolute chronology.	Compare new remote sensing data sets with Apollo-era data sets to detect formation of new craters.	Higher-resolution images to provide targeted crater counts of undisturbed ejecta surfaces from, e.g., Orientale, Imbrium, and Copernicus.	Sample return from key benchmark basins, craters (e.g., Copernicus, Tycho), and lava flows for precise isotopic dating.	Develop instruments for precise, in situ geochronology. Use landed geochemical instrumentation to identify best samples for return.	Human geologic fieldwork increases chances of recognizing the best samples.

Goal					
4a. Determine the compositional state (elemental, isotopic, mineralogic) and compositional distribution (lateral and depth) of the volatile component in lunar polar regions.	Analyze existing and new data and integrate information (photometry, morphology, topography, temperature, hydrogen and surface frost distribution) from the polar regions to improve knowledge of volatile spatial distribution.	High-spatial resolution distribution of volatiles on and in the regolith poleward of 70 degrees.	Cryogenically preserved samples to determine the detailed elemental and isotopic composition of soils from permanently shaded regions.	Measure elemental and isotopic composition of gas evolved from regolith in permanent shade heated up to 700 K, obtained from depths greater than 10 cm and up to a meter. Determine the presence of refractory volatile-bearing species including water-bearing minerals, complex organics, and clathrates. Determine elemental composition, especially hydrogen, for immediate surroundings of sampling site. Determine any local stratigraphy using geophysical methods. Support technology development for operation at low temperatures for long duration.	Search for evidence of complex volatile history of polar soils, including examination and sampling of shallow and deep trenches. Support technology development for field studies in regions of permanent shade, including aspects of site disturbance by high-temperature equipment.
3a. Determine the extent and composition of the primary feldspathic crust, KREEP layer, and other products of planetary differentiation.	High-resolution global maps of mineralogy and geochemistry to characterize important geological and geochemical units.	Geophysical measurements of representative regions. Higher-resolution geophysical (e.g., seismic, gravitational) measurements for modeling crustal structure and understanding extent of crustal units.	Sample return from major lunar terranes (e.g., feldspathic highlands, SPA Basin, PKT) for detailed geochemical and isotopic analysis.	Establish/participate in an international network of ground-based geophysical instruments. Determine mineralogy and petrology across multi-kilometer traverses at selected craters.	Acquire wider variety of samples (for in situ study and return) through sifting, drilling, and identification of unusual samples.
2a. Determine the thickness of the lunar crust (upper and lower) and characterize its lateral variability on regional and global scales.	Lateral variability addressed by careful analysis of high-quality topography (LRO) with existing gravity models.	High-resolution gravity measurements to allow separation of flexure, density, and thickness effects.	Samples from previously unsampled regions to provide constraints on lateral variability of crustal composition and density.	Establish a geophysical network to determine crustal thickness from analyses of natural and artificial seismic events.	Human installation of seismic instrumentation provides for better sensitivity to extremely small lunar seismic signals.

TABLE 5.1 Continued

Science Goals	(a) **Information Extraction** An enabling new framework for lunar exploration will be provided by data from SMART-1, SELENE, Chang'e, Chandrayaan-1, and LRO. The assumption is that all missions and key instruments will be successful.	(b) **Orbital Measurements** Orbital measurements are not included in the complement of missions planned for launch by 2008. The assumption is that the four missions planned will return appropriate data as planned; if not, new measurements that provide similar high-priority compositional and geophysical data need to be acquired.	(c) **Sample Return** The types of returned samples and of science analyses required are identified.	(d) **Landed Experiments, Instruments, and Rovers** These include science measurements for/from landed sites; category also encompasses penetrators/impactors.	(e) **Human Fieldwork** Science areas that specifically benefit from human capabilities are identified.
2b. Characterize the chemical/physical stratification in the mantle, particularly the nature of the putative 500-km discontinuity and the composition of the lower mantle.			Return of samples representative of the upper mantle (e.g., SPA Basin) provides a geochemical framework for the interpretation of seismic and magnetic sounding data.	Establish a global (including farside) geophysical network to determine the seismic and resistivity structure of the mantle from analyses of moonquake signals and low-frequency electromagnetic sounding.	Human installation of seismic instrumentation provides for better sensitivity to extremely small lunar seismic signals.
8a. Determine the global density, composition, and time variability of the fragile lunar atmosphere before it is perturbed by further human activity.	Ultraviolet spectral measurements from LRO will provide upper limits or measurements of species such as Ar using resonantly scattered sunlight.	Mass spectrometers flown in low lunar orbit (<50 km) could provide in situ measurements of lunar atmospheric species.		A network of surface mass spectrometers could monitor poleward migration of volatiles.	Backpack mass spectrometers could provide sensitive monitoring of atmospheric pollution.

2c. Determine the size, composition, and state (solid/liquid) of the core of the Moon.				Establish a global (including farside) geophysical network to determine the seismic and resistivity structure of the deep interior from analyses of moonquake signals and low-frequency electromagnetic sounding, and to improve the measurement of the dynamical parameters of the Moon through Earth-based laser tracking.	Human installation of seismic instrumentation provides for better sensitivity to extremely small lunar seismic signals.
3b. Inventory the variety, age, distribution, and origin of lunar rock types.	Continue to search for exotic components in existing samples and remote sensing data. Use remote sensing to help determine the regional geologic context of returned samples.	Higher-resolution global and regional mineralogic and geochemical maps to identify unusual lithologies and provide context for returned samples and meteorites. Higher-resolution images to provide targeted crater counts.	Sample returns from representative and previously unsampled locations (selected from global compositional maps).	Develop instruments for precise, in situ geochronology. Use landed geochemical instrumentation to identify best samples for return.	Acquire wider variety of samples (for in situ study and return) through sifting, drilling, identification of unusual samples.
8b. Determine the size, charge, and spatial distribution of electrostatically transported dust grains and assess their likely effects on lunar exploration and lunar-based astronomy.	LRO cameras may detect horizon glow during limb-scanning operations.	Dedicated limb-scanning measurements of scattered sunlight from dust clouds could provide maps of lunar dust transport.		Deposition of lunar dust on optical surfaces could be monitored by surface-based instruments.	Determine effects of human activity on dust environment.

NOTE: Acronyms are defined in Appendix B.

PRIORITIZATION OF SCIENCE CONCEPTS

The eight science concepts discussed in Chapter 3 address broad areas of scientific research. Each has multiple components and is linked (see Table 3.1) to different aspects of the overarching themes—early Earth-Moon system, terrestrial planet differentiation and evolution, solar system impact record, and lunar environment—presented in Chapter 1. In addition, there are multiple avenues for implementation (information extraction; orbital measurements, sample return; landed experiments, instruments, and rovers; and human fieldwork; see Table 4.1). In order to provide a sense of the overall importance of each of the eight broad science concepts, the committee evaluated only the scientific merit of each concept to rank order these concepts. They are listed in order throughout this report and in Tables 3.1 and 4.1. It should be noted that *all* concepts discussed are viewed to be scientifically important and their ordering in this report is simply a relative ranking:

1. The bombardment history of the inner solar system is uniquely revealed on the Moon.
2. The structure and composition of the lunar interior provides fundamental information on the evolution of a differentiated planetary body.
3. Key planetary processes are manifested in the diversity of lunar crustal rocks.
4. The lunar poles are special environments that may bear witness to the volatile flux over the latter part of solar system history.
5. Lunar volcanism provides a window into the thermal and compositional evolution of the Moon.
6. The Moon is an accessible laboratory for studying the impact process on planetary scales.
7. The Moon is a natural laboratory for regolith processes and weathering on anhydrous airless bodies.
8. Processes involved with the atmosphere and dust environment of the Moon are accessible for scientific study while the environment remains in a pristine state.

PRIORITIZATION OF SCIENCE GOALS

Within the science concepts, the committee identified 35 specific science goals that can be addressed at least in part during the early phases of the VSE. For these science goals, the committee evaluated their science merit as well as the degree to which they are possible to achieve within the limits of current or near-term technical readiness and practical accessibility. Within their respective science concepts, these goals are listed in order of their overall priority ranking (a-e) in Table 3.1.

The committee also evaluated and rank ordered all 35 specific science goals together, apart from the science concepts with which they are grouped. The 11 highest-ranking lunar science goals are listed below and in Table 5.1 in priority order. To achieve this group of goals, the committee identified possible means of implementation (see Table 5.1).

The committee's highest-priority science goals are the following:

• *1a. Test the cataclysm hypothesis by determining the spacing in time of the creation of the lunar basins.* The history of impacts in the early Earth-Moon system, in particular around 3.9 Ga, the time that life was emerging on Earth, is a critical chapter in terrestrial planet evolution. Understanding this period is important for several reasons: as tests of our models of the impact rate, planetary accretion, impact frustration of life, magma ocean formation and evolution, and extension and verification of the chronology. In order to answer the question of whether there was a cataclysm at 3.9 Ga, sample returns from the oldest impact basins combined with high-resolution imaging from orbit are required.

• *1b. Anchor the early Earth-Moon impact flux curve by determining the age of the oldest lunar basin (South Pole-Aitken Basin).* Although the enormous South Pole-Aitken Basin is stratigraphically the oldest basin on the Moon, its absolute age is completely unconstrained. All models of the first few hundred million years of solar system history depend on whether the large basins are part of a decreasing flux of material swept up by growing planet embryos or a later separate pulse of planetesimal-sized bodies. Details of the lunar stratigraphy can be better defined by integrated high-resolution imagery and topography, but it is essential to provide an absolute date for the oldest basin, the South Pole-Aitken Basin, with the type of precision that can only be obtained in Earth-based laboratories with returned samples.

- *1c. Establish a precise absolute chronology.* A well-calibrated lunar chronology not only can be used to date unsampled lunar regions, but it can also be applied to date planetary surfaces of other planets in the inner solar system through modeling. An absolute lunar chronology is derived from combining lunar crater counts with radiometric sample ages and is thus the most precise—and in some cases the only—technique to date planetary surfaces for which samples have not been or cannot be obtained. In order to determine the precise shape of the lunar chronology curve, samples should be returned from several key benchmark craters, young lava flows, and old impact basins, which also need to be imaged at high spatial resolution.

- *4a. Determine the compositional state (elemental, isotopic, mineralogic) and compositional distribution (lateral and depth) of the volatile component in lunar polar regions.* The extremely low temperature surfaces in permanent shade at the lunar poles have been accumulating ices and other volatile-bearing materials for at least 2 billion years. This potential scientific bonanza contains information on the history of volatile flux in the recent solar system and is a natural laboratory for studying how volatiles develop in the space environment. However, there is a near-total lack of understanding of the nature and extent of these polar materials. Landed missions to the poles will produce entirely new knowledge of this unknown territory.

- *3a. Determine the extent and composition of the primary feldspathic crust, KREEP layer, and other products of planetary differentiation.* The lunar magma ocean has been the cornerstone of lunar petrology since the return of the Apollo samples and has gone on to form the basis for our understanding of differentiation processes in all the terrestrial planets, including Earth and Mars. Many details of the differentiation process can be told through the geochemistry and distribution of key lunar rock types that we think are primary products. Regional orbital remote sensing will be needed to identify areas that contain these rocks and how they fit into the global picture. Geophysical data, particularly seismic profiling of the lunar crust, help identify the depth and extent of important layers in the lunar crust. Both human and robotic landed missions can provide targeted sample return so that we can study these products in the same detail as for the Apollo samples.

- *2a. Determine the thickness of the lunar crust (upper and lower) and characterize its lateral variability on regional and global scales.* The lunar crust provides basic constraints on the characteristics of the lunar magma ocean from which it formed. Its volume fixes the extent of differentiation of the original lunar material, and differences between the upper and lower crust, along with global-scale variations in thickness, provide essential clues to the processes that formed the outermost portions of the Moon. A seismic network of at least regional extent is essential for providing this information.

- *2b. Characterize the chemical/physical stratification in the mantle, particularly the nature of the putative 500-km discontinuity and the composition of the lower mantle.* The structure of the mantle has been affected by the initial differentiation of the Moon by magma ocean fractionation and core formation as well as any subsequent evolution, such as mantle overturn and sub-solidus convection. All of these processes will have left their marks in terms of compositional and mineralogical stratification, and detailed knowledge of this structure may allow us to decipher the Moon's earliest history. The seismic discontinuity tentatively identified by the Apollo seismic experiment has particular significance in differentiation models, as it may represent the base of the original magma ocean. The only effective methods for probing the lunar mantle are global-scale seismology and electromagnetic sounding.

- *8a. Determine the global density, composition, and time variability of the fragile lunar atmosphere before it is perturbed by further human activity.* Although the density of the lunar atmosphere was fairly well characterized on the nightside by Apollo, 90 percent of the atmospheric constituents were not identified. The measurements need to be extended to the dayside, and the composition of the atmosphere should be determined as completely as possible. It is crucial that these measurements be made before the atmosphere is perturbed by future human landings; therefore, this topic was ranked higher in implementation priority than in scientific priority. Both orbital and surface deployments of mass spectrometers are needed to make the required measurements.

- *2c. Determine the size, composition, and state (solid/liquid) of the core of the Moon.* At this point the very existence of a metallic lunar core, while likely, has not been fully established. Yet its size and composition play a fundamental role in determining the initial bulk composition of the Moon and the subsequent differentiation of the mantle, as well as the Moon's thermal and magnetic history. Measurements from a globally distributed network of seismometers, augmented by electromagnetic sounding and precision laser tracking of variations in lunar rotation, will be necessary to characterize the lunar core.

• *3b. Inventory the variety, age, distribution, and origin of lunar rock types.* After the formation of the primary products of the lunar magma ocean, the Moon continued to produce a rich diversity of rocks by numerous geologic processes. Erupted basalts, emplaced plutons, and remelted impact-melt sheets all contain clues to continued geologic activity on the Moon and the processes that enabled this activity. Understanding when and how the diversity of lunar rocks formed and how they are at present distributed allows the prediction of where else on the Moon they may be located, even if they are not expressed at the surface. Laboratory analysis of returned samples from diverse locations on the Moon enables complete, high-precision geochemical, mineralogical, and isotopic characterization of diverse lunar rocks. Higher-resolution geochemical and mineralogical remote sensing databases are also crucial in providing geologic context for unusual lithologies.

• *8b. Determine the size, charge, and spatial distribution of electrostatically transported dust grains and assess their likely effects on lunar exploration and lunar-based astronomy.* Lunar dust is an important constituent of the lunar environment. Because of illumination by sunlight and the impact of the solar wind, the dust is electrostatically charged and is levitated and transported by electric fields produced by the solar wind. The transport of the dust and its deposition on surfaces will place important limitations on human activities and on astronomical observations that may be planned for the Moon. Surface measurements of dust, which can be made robotically and later with astronaut assistance, are needed to characterize the dust environment and its effects on deployed systems and instrumentation.

INTEGRATED HIGH-PRIORITY FINDINGS AND RECOMMENDATIONS

In arriving at the priority science concepts presented in Chapter 3 and the specific goals presented in Chapter 3 and above, the committee found that there were a number of larger integrated issues and concerns that were not fully captured either in the discussion of the science concepts or in the science goal priorities and their implementation. The committee therefore developed a group of integrating findings and recommendations that envelop, complement, and supplement the scientific priorities discussed in the report:

Finding 1: Enabling activities are critical in the near term.

A deluge of spectacular new data about the Moon will come from four sophisticated orbital missions to be launched between 2007 and 2008: SELENE (Japan), Chang'e (China), Chandrayaan-1 (India), and the Lunar Reconnaissance Orbiter (United States). Scientific results from these missions, integrated with new analyses of existing data and samples, will provide the enabling framework for implementing the VSE's lunar activities. However, NASA and the scientific community are currently underequipped to harvest these data and produce meaningful information. For example, the lunar science community assembled at the height of the Apollo program of the late 1960s and early 1970s has since been depleted in terms of its numbers and expertise base.

Recommendation 1a: NASA should make a strategic commitment to stimulate lunar research and engage the broad scientific community[2] by establishing two enabling programs, one for fundamental lunar research and one for lunar data analysis. Information from these two recommended efforts—a Lunar Fundamental Research Program and a Lunar Data Analysis Program—would speed and revolutionize understanding of the Moon as the Vision for Space Exploration proceeds.

Recommendation 1b: The suite of experiments being carried by orbital missions in development will provide essential data for science and for human exploration. NASA should be prepared to recover data lost due to

---

[2]See also National Research Council, *Building a Better NASA Workforce: Meeting the Workforce Needs for the National Vision for Space Exploration*, The National Academies Press, Washington, D.C., 2007.

failure of missions or instruments by reflying those missions or instruments where those data are deemed essential for scientific progress.

Finding 2: Strong ties with international programs are essential.

The current level of planned and proposed activity indicates that almost every space-faring nation is interested in establishing a foothold on the Moon. Although these international thrusts are tightly coupled to technology development and exploration interests, science will be a primary immediate beneficiary. NASA has the opportunity to provide leadership in this activity, an endeavor that will remain highly international in scope.

Recommendation 2: NASA should explicitly plan and carry out activities with the international community for scientific exploration of the Moon in a coordinated and cooperative manner. The committee endorses the concept of international activities as exemplified by the recent "Lunar Beijing Declaration" of the 8th ILEWG (International Lunar Exploration Working Group) International Conference on Exploration and Utilization of the Moon (see Appendix D).

Finding 3: Exploration of the South Pole-Aitken Basin remains a priority.

The answer to several high-priority science questions identified can be found within the South Pole-Aitken Basin, the oldest and deepest observed impact structure on the Moon and the largest in the solar system. Within it lie samples of the lower crust and possibly the lunar mantle, along with answers to questions on crater and basin formation, lateral and vertical compositional diversity, lunar chronology, and the timing of major impacts in the early solar system.

Missions to South Pole-Aitken Basin, beginning with robotic sample returns and continuing with robotic and human exploration, have the potential to be a cornerstone for lunar and solar system research. (A South Pole-Aitken Basin sample-return mission was listed as a high priority in the 2003 NRC decadal survey report *New Frontiers in the Solar System: An Integrated Exploration Strategy*.[3])

Recommendation 3: NASA should develop plans and options to accomplish the scientific goals set out in the high-priority recommendation in the National Research Council's *New Frontiers in the Solar System: An Integrated Exploration Strategy*'s (2003) through single or multiple missions that increase understanding of the South Pole-Aitken Basin and by extension all of the terrestrial planets in our solar system (including the timing and character of the late heavy bombardment).

Finding 4: Diversity of lunar samples is required for major advances.

Laboratory analyses of returned samples provide a unique perspective based on scale, precision, and flexibility of analysis and have permanence and ready accessibility. The lunar samples returned during the Apollo and Luna missions dramatically changed understanding of the character and evolution of the solar system. Scientists now understand, however, that these samples are not representative of the larger Moon and do not provide sufficient detail and breadth to address the fundamental science concepts outlined in Table 3.1 in this report.

Recommendation 4: Landing sites should be selected that can fill in the gaps in diversity of lunar samples. Mission plans for each human landing should include the collection and return of at least 100 kg of rocks from diverse locations within the landing region. For all missions, robotic and human, to improve the probability of finding new, ejecta-derived diversity among smaller rock fragments, every landed mission that will return to Earth should retrieve at least 1 kg of rock fragments 2 to 6 mm in diameter separated from bulk soil. Each mission should also return 100 to 200 grams of unfractionated regolith.

---

[3]National Research Council, *New Frontiers in the Solar System: An Integrated Exploration Strategy*, The National Academies Press, Washington, D.C., 2003.

6

Observations and Science Potentially Enabled by the Vision for Space Exploration

The Vision for Space Exploration (VSE) and the returns to the Moon by robotic and human explorers will provide the opportunity to make use of the Moon's location and the unique perspective that this location provides to carry out research in several fields of science. In this chapter the committee considers the potential benefits for various disciplines—astrophysics, gravitational physics, cosmic-ray physics, astrobiology, earth sciences, heliophysics, and magnetospheric and ionospheric physics—from of emplacing experimental apparatus on the Moon.

This committee was not constituted to recommend priorities for research in these diverse disciplines. The committee believes that the appropriate standing committees of the Space Studies Board are better able to judge the scientific priorities of research within their disciplines in the context of such research conducted from other locations in space, as well as the Moon.

ASTRONOMY AND ASTROPHYSICS

The lunar surface has historically been seen as a possible site for telescopes operating panchromatically and at the highest resolution, without an intervening absorbing and distorting atmosphere. In the early years of space astronomy, the stability of the lunar surface and large reaction mass were considered highly advantageous for telescope pointing and tracking, and the long diurnal cycle was considered advantageous, at least relative to low-Earth orbit (LEO), for thermal equilibration of large optical instruments. In addition, hypothetical future human visits to the Moon ensured some level of accessibility for service, maintenance, and upgrades.

The full costs of construction, deployment, and operation of telescopes on the lunar surface can reasonably be assumed to be larger than the costs for comparable equipment operating in free space, however. This cost difference arises, at minimum, from added propulsion requirements and the mitigation of added risks from landings. The issue then becomes a question of what features the lunar surface uniquely offers astronomical telescopes that free space does not.

The lunar surface offers, at least in the early stages of lunar development, extraordinarily radio-quiet sites on the lunar farside that could enable a highly sensitive low-frequency radio telescope.[1] Such a telescope would be a

---

[1]The International Telecommunication Union (ITU), through its Radiocommunication Sector, has formally recognized the scientific importance of lunar radio astronomy, especially below 30 MHz, in its recommendation ITU-RA.479 "Protection of frequencies for radioastronomical measurements in the shielded zone of the Moon." The committee notes, however, that that the low level of radio interference in this zone could be compromised by noncompliant development on the farside surface, telecommunication and research spacecraft in lunar orbit (including reflections), or science missions elsewhere, notably at the Earth-Sun Lagrange points.

powerful tool in investigating the "dark ages" of the universe, before the reionization era, in which highly redshifted 21 cm (1420 MHz) emission from neutral hydrogen would reveal the earliest structures in the universe before the first phase of nuclear enrichment. At redshifts on the order of one hundred, this strong emission line approaches the ~15 MHz plasma frequency of Earth's ionosphere, below which the opacity is very high and observations from the surface of Earth are not possible. For redshifts below this, the science thrust is for terrestrial efforts (e.g., Low Frequency Array, Square Kilometer Array), although such implementations are handicapped on Earth compared with the lunar farside, because of the severe human-produced and ionospheric radio-frequency noise. These low frequencies, unobservable from Earth, have special relevance also for heliophysical research on solar bursts and particle acceleration processes within them. A low-frequency radio interferometer with simple dipole elements spread out on a kilometer scale of the lunar surface thus has synergistic value to both priority astrophysical and heliophysical research. As a result of this synergy, as well as the usefulness of such low frequencies to understanding particle acceleration mechanisms in active galaxies, a modest nearside array has substantial scientific value both for these bright astronomical sources and also for allowing the refinement of technology and understanding of environmental issues.

An innovative concept recently proposed would have a complete antenna line electrodeposited on a long strip of polyamide film. In this form, a lightweight array could be simply unrolled onto the lunar surface in one of the first lunar return trips (see Figure 6.1). Unlike efforts in the optical, infrared (IR), or ultraviolet (UV) domains,

FIGURE 6.1 An artist's conception of the Radio Observatory for Lunar Sortie Science farside Mylar dipole radio telescope (shown in yellow); an active Sun is shown on the upper right, and Earth's magnetosphere is shown at the upper left. SOURCE: Courtesy of J. Lazio, Naval Research Laboratory; R. MacDowall, NASA/Goddard Space Flight Center; J. Burns, University of Colorado; D. Jones, Jet Propulsion Laboratory; K. Weiler, Naval Research Laboratory; and J. Kasper, Massachusetts Institute of Technology.

such a telescope would not be affected by dust deposition. As a result of the high astronomical priority of this work and the uniquely enabling character of the radio-quiet farside lunar surface, such efforts deserve cultivation.

With this in mind, near-term studies should be started to improve the understanding of the requirements and possible limitations of such a low-frequency radio interferometer effort, perhaps defining near-term site survey experiments that would help clarify the potential. What are the optimal sites for such an installation? There is also need to extend the original 40-year-old work that identified the "quiet zone of the Moon" and which is the only source of data on this subject to this day. What are the temporal characteristics of the frequency environment? To the extent that wide access to the lunar surface is not provided in the current VSE architecture, how close can such a facility be to the planned polar outpost and still have optimal performance? Given that farside human landings are not currently planned, how would such a facility be deployed and operated? Can a credible design for a major installation be developed that does not assume pre-existing infrastructure (communications, power, and so on)? With the recent predictions that sunlit/shaded edges of the Moon will develop substantial electrical potential gradients, are electrostatic discharges a possible natural noise source?

Currently, conventional single-aperture, single-spacecraft telescopes in free space that do not require precision constellation management now use proven pointing and tracking technology. The platform of the Moon is by no means as enabling for astronomy as it was once thought to be. Lunar gravity, while small compared with that on Earth, nevertheless requires that telescopes on the Moon be more massive than those based in free space in order to achieve a requisite stiffness that can preserve optical alignment as the telescope moves across the sky. However, a large telescope can be expected to have its surface shape maintained by active optics, and the issues of maintaining the surface by inertial reaction to the structure and the problem of free vibrations of that structure become larger as the telescope becomes larger. In the range of telescopes as large as the James Webb Space Telescope (JWST), such considerations seem to favor free-space locations. For larger apertures these issues will need to be re-examined. The frigid conditions in lunar polar craters might serve the needs of future thermal infrared telescopes, but passive cooling strategies now being designed into JWST for Earth-Sun Lagrange point L2 provide such low temperatures at modest fractional cost.

For ultraviolet astronomy and for astronomy involving the precise control of scattered background emission (e.g., planet detection), tolerance for dust contamination is very small. For thermal infrared astronomy, the observational background (which determines the sensitivity) is proportional to the absorbtivity (and re-radiation), not the reflectivity, so small dust deposits on the optics can seriously compromise the performance. As a result, the lunar surface appears to be minimally suitable for large UV/optical/IR telescopes.

Accessibility by humans is not clearly a unique advantage offered by the lunar surface. There is a large base of experience in deploying, maintaining, and servicing large telescopes in free space from the Hubble Space Telescope, as well as substantial expertise in in-space construction of large facilities from the International Space Station experience. The lunar-return implementation architecture for the Vision for Space Exploration, now in development, does not provide explicitly for in-space opportunities. Nevertheless, it appears likely that telescopes in free space, whether in LEO or at the dynamically convenient and thermally highly optimal Earth-Sun Lagrange points, may eventually be serviceable by humans and robots using augmentations to this architecture. In the case of remote Earth-Sun Lagrange points, such access could most conveniently be ensured after returning these telescopes by known pathways that require only small changes in spacecraft velocity to get to closer sites, such as the Earth-Moon Lagrange points. In this context, however, the lunar surface may well play a key role in operational support for such relocation efforts.[2]

The lunar surface gravity and solid surface do offer potential advantages for particular lunar observatory architectures and deserve further consideration by way of trade-off studies against functionally comparable free-space facilities. Several concepts may be viable, at least in the long term. A large liquid mirror telescope on the lunar surface near a pole that could offer extremely deep observations into very limited regions of the sky near the local

---

[2]The development of a heavy-lift (Ares 5) launcher could offer major advantages to astronomy, providing the ability to lift large telescopes, fully assembled, into free space. In these respects, the astronomy community is particularly excited about the space transportation capabilities that will arise as the results of the lunar exploration architecture. See, for example, the proceedings of the Astrophysics from the Moon conference, Space Telescope Science Institute, 2007, at www.stsci.edu/institute/conference/moon, and the Lunar Exploration Science Workshop, Tempe, Arizona, February 27-March 2, 2007, at https://www.infonetic.com/tis/lea/, accessed May 29, 2007.

zenith has been proposed. While highly innovative, the concept is technologically challenging, requiring a liquid with very low freezing temperature (<100 K) that can be flash coated, as well as a large precision-rotating tray for it. The very small field of view thus far proposed in the optical designs is, however, a significant handicap in meeting current National Research Council decadal survey science priorities.[3] The surface also offers, as a solid optical bench, one strategy for deploying a multi-telescope, optically coupled interferometer that would offer high spatial resolution for astronomical sources. Such a design follows the pattern of terrestrial interferometers. However, a free-flying array in space could acquire data from a wider range of spatial frequencies than the range accessible to an interferometer fixed on the Moon. While lunar ultraviolet, optical, or infrared interferometers might be achievable with technology similar to that being used on Earth, the telescope emplacement, precise linkage, and sophisticated fringe tracking are not easy, and technology developments providing capability in free space would offer more potential. To the extent that such telescopes might someday be considered to be truly viable for priority astronomy, a careful assessment of dust contamination, both from natural processes and from human operations on the surface, will be necessary.

Finally, can the very substantial computing requirements of a future array for cosmological studies be credibly powered at a remote lunar site that is unlikely to see continuous solar illumination, especially in view of the fact that the quietest radio-frequency conditions will be at night? In addition, serious consideration needs to be given to the preservation of low radio noise on the farside. Are there strategies that can offer high-performance lunar global communication (e.g., laser communications) that will minimally compromise the scientific resource on the Moon as the Moon is developed?

Recent creative suggestions for astronomical instrumentation in lunar orbit deserve special mention. With no atmosphere to blur the horizon, sensor payloads in lunar orbit can use the edge of the Moon as an occulting knife edge for the precise localization of high-energy photons. With a detector spacecraft in lunar orbit, beam cutoff timing provides a source position in one direction. With apsidal plane motion of the orbit, many parts of the sky can be examined in different position angles, allowing a small spacecraft in lunar orbit with only modest pointing control to achieve very high angular resolutions for astronomical sources. The edge of the Moon has also been suggested as a convenient occultation strip to block out the Sun while doing deep infrared imaging in its vicinity in order to measure zodiacal dust at <1 astronomical unit (AU). Such instrument packages may be quite small and might find early use as attached payloads on future lunar surveillance orbiters, as well as on Crew Exploration Vehicle service modules that are parked in lunar orbit.

Other novel ideas for astronomical telescopes have been proposed in which the lunar regolith itself can be used as a high-energy particle and gravitational wave detector. While the lunar regolith is potentially enabling in this regard, the experiment would require substantial civil engineering in the form of excavation, drilling, and the movement of large volumes of regolith.

In the near term, opportunities for very small, "suitcase astronomy" (<100 kg) payloads, comanifested with equipment on an LSAM, might offer some astronomical value. A particularly potent example that does not even require such external supply equipment is evident in laser ranging reflectors. Such reflectors, as deployed on the Moon in the Apollo program, provided important data on the lunar orbit and (see Box 6.1) are now challenging understanding of basic physics. Other comanifested instruments might use power and communication from the Lunar Surface Access Module or outpost and be self-sufficient with regard to pointing and tracking. Although the committee is not aware of any such instrumentation that would address decadal survey priority astronomy (because the collecting areas are necessarily very small), a "free ride" to a lunar platform with a very modest, competitively selected instrument might offer science value. The opportunity presented by such small instruments as attached lunar payloads will have to be carefully evaluated with respect both to conventional implementation in free space (e.g., NASA's Small Explorer (SMEX)-Lite and University Explorer missions) and to attachment to small missions at the Moon-Earth Lagrange point L1, or in lunar orbit.[4]

---

[3]National Research Council, *Astronomy and Astrophysics in the New Millennium*, National Academy Press, Washington, D.C., 2001.

[4]It should be understood that while lunar development in the long run (perhaps including in situ resource utilization [ISRU] strategies) provides real cost offsets to telescope development, the cost/value relationships for lunar telescopes may change. While many factors suggest that free space may offer the highest astronomical performance in the near future for telescopes in space, cost offsets may derive in the more distant future as a result of lunar surface activities with other goals in mind (e.g., ISRU).

BOX 6.1
Lunar Laser Ranging: An Example of an Enabled Experiment

Lunar laser ranging (LLR) measures the round-trip travel time of short laser pulses that are reflected back to Earth from corner-cube (retroreflector) arrays on the Moon. The range data are used to perform a general phenomenological check on our understanding of gravity—the weakest of the fundamental forces of nature. They also enable the refinement of the constants in the parameterized post-Newtonian formulation of general relativity and the testing of other theories of gravitation. LLR has provided the most precise limits to date on the following properties of gravity:

- Weak equivalence principle differential (fractional) acceleration $\Delta a/a < 1.4 \times 10^{-13}$,
- Strong equivalence principle (self gravity) to $<4.5 \times 10^{-4}$,
- Time-rate-of-change of Newton's gravitational constant to $\dot{G}/G < 10^{-12}$ per year,
- Gravitomagnetism (frame dragging) to 0.1 percent,
- Geodetic precession to 0.6 percent, and
- Inverse square law to $<10^{-10}$ times the strength of gravity at 10^8 m length scales.

The first placement of retroreflectors on the lunar surface was by the Apollo astronauts in 1969. Recently, the Apache Point Observatory Lunar Laser-ranging Operation (APOLLO) began achieving 1 mm precision on lunar range. Deployment of next-generation reflectors would allow range precision to approach the 0.1 mm level, a better than two orders-of-magnitude improvement over the data used to determine the limits on the gravitational properties listed above.

Placement of a single new array would be helpful, although by itself incapable of adequately constraining lunar rotational and tidal motions.

Placement of three total new arrays widely distributed near the limb would optimally benefit high-precision ranging. The distribution of such reflectors/transponders may naturally and economically be associated with the implementation of a geophysical network.

In addition to probing the predictions of gravitational physics, the rich LLR data set can be used for other scientific purposes. For example, by monitoring the physical reactions to known torques on the Moon, the properties of the lunar interior, such as the presence of a liquid core and the interaction between this core and the surrounding mantle, may be discerned. The LLR data series can also be used to study the orientation of Earth in space, which connects to climate monitoring, and geophysics. Solar system navigation also benefits from the LLR data series.

SOURCE: Adapted from T.W. Murphy, University of California, San Diego, "Lunar Reflectors for Tests of Relativity," white paper for the committee, 2006.

There are near-term scientific investigations that would clarify some of the issues noted above. Of primary interest are studies of lunar dust resulting in a clear understanding of the threat that dust produced by exploration activity and natural processes poses to optical and mechanical systems and how dust enhances sky backgrounds, as well as strategies for mitigating this threat. Also needed is long-term characterization of the natural seismic environment, to be complemented by predictions of the induced seismic environment caused by future human lunar operations. A careful assessment of radio noise on the lunar farside, which may see contributions from local electrostatic discharges, is also of importance.

Thus, there are synergistic opportunities for combining site testing for astronomy with studies related to Earth observations and research on the lunar dust and plasma environment to yield data of interest to several scientific disciplines.

ASTROBIOLOGY

Astrobiology concerns the origin and development of life, the existence of life elsewhere, and the future of life on Earth and in the universe. There is great overlap between the aims of astrobiology and the disciplines of astronomical, planetary, geological, and biological sciences; the significance of astrobiology comes from the integration of these areas. The origin and evolution of planets, the development of planetary crusts, and the impact and radiation environment that planetary bodies experience are critical boundary conditions for the habitability of Earth and other planets. A report by the NASA Astrobiology Institute[5] outlines the relevance of the Moon to astrobiology, and many of the goals and objectives of the Astrobiology Roadmap[6] are addressed by investigations recommended throughout this report.

The nature of the Moon and its accessibility provide a unique opportunity to make significant advances in several aspects of astrobiology. The importance of understanding the evolution of the lunar crust and other aspects of the Moon as a planetary end member is well recognized. Two specific areas have been called out for particular attention in the lunar context: (1) the early and recent impact flux and (2) the history of the lunar (and hence terrestrial) radiation environment.

The Moon provides a unique environment with which to study the early and late impact flux at 1 AU. The importance and influence of large, basin-sized impacts on the early evolution of life remain unresolved, and whether Earth experienced a late spike in basin-scale impacts (the terminal cataclysm) or a rate more smoothly declining is a critical boundary condition in this debate. Thus, the finding in this report that a South Pole-Aitken Basin sample return can make a vital contribution to the existence of a cataclysm would result in a major contribution to a high-priority astrobiological goal. However, the recent impact flux, well preserved and exposed on the Moon, is relevant to the development of life over the Phanerozoic, as impacts are a strong candidate for the cause of several major extinctions. Absolute age dating of stratigraphically important recent craters and a statistical sample of dated smaller craters would definitely determine the recent flux constraining the influence of impact on the evolution of life (extinction and evolutionary radiation).

The second identified priority for lunar astrobiological investigations is that of understanding the past energetic particle and plasma environment of the terrestrial planets. The particle flux is a critical unknown in understanding the ancient habitability of Earth. The early Sun was expected to have had a more intense solar wind and solar flare flux at early times, while the solar system has been repeatedly exposed to supernova and gamma-ray burst events and experiences a variable cosmic-ray flux as the Sun travels through the interstellar medium. Ancient lunar regolith recorded this environment in the form of trapped gases, fission products, and tracks. Sampling and analysis of ancient soil trapped between datable lava flows is the only known way to study the special circumstances at that critical time. While such sampling may be ambitious for an early robotic program, candidate sites can be identified and thoroughly characterized using high-resolution imaging and remote sensing for later detailed study.

Additional areas recently identified to be of high scientific interest to astrobiology include (1) development of understanding of organic chemistry in the lunar cold traps, (2) characterization of extrasolar planetary bodies with large lunar telescopes, and (3) search for possible components of ancient Earth in lunar regolith. The latter two are linked to the long-term potential of using the Moon for science as VSE infrastructure and facilities are emplaced.

Evaluating and sampling materials from the unusual environment at the lunar poles have been identified as a lunar science goal and are discussed in Chapters 3 through 5 of this report. As a potential repository of solar system volatiles and organics, polar material feeds directly into astrobiology concerns. The astrobiological significance of polar samples places a greater emphasis on cryogenic sample return, or alternatively on an appropriately equipped laboratory facility at a polar outpost supported by human fieldwork.

In identifying and characterizing extrasolar planets, there is congruence of the objectives of astronomy and astrophysics with those of astrobiology-related observational goals, namely, to maximize the effective aperture of telescopes. In the near term, tasks related to understanding the lunar environment (dust mobility, seismic stability, and so on) that enable such facilities to be evaluated and planned apply to astrobiology concerns as well as to astronomy and astrophysics. Earth-shine observations have been used to observe Earth's spectrum near full phase,

---

[5]B. Jakosky, A. Anbar, G.J. Taylor, and P. Lucey, "Astrobiology Science Goals and Lunar Exploration: NASA Astrobiology Institute White Paper," 2004.

[6]D.J. DesMarais, L.J. Allamandola, S.A. Benner, A.P. Boss, J.R. Cronin, D. Deamer, P.G. Falkowski, J.D. Farmer, S.B. Hedges, B.M. Jakosky, A.H. Knoll, D.R. Liskowsky, V.S. Meadows, M.A. Meyer, C.B. Pilcher, K.H. Nealson, A.M. Spormann, J.D. Trent, W.W. Turner, N.J. Woolf, and H.W. Yorke, "The astrobiology roadmap," *Astrobiology* 3(2):219-233, 2003.

but they fail to provide phase information, nor do they allow polarimetric studies such as those needed to identify liquid water. In preparation for the study of extrasolar planets, there could be particular value in making synoptic observations of Earth. For example, the vantage from the Moon (or a similar orbit) could allow the monitoring of specific properties of Earth as it would appear as a single point object, going through phases. Such data might identify readily measurable properties that are diagnostic of our habitable world, such as looking for possible evidence of oceans from the polarization of Sun glint and of vegetation from any cyclical spatial variation of the strong chlorophyll absorption edge in the visible to near-infrared spectrum.

We know that small amounts of material are redistributed from one body to another by impacts (e.g., meteorites), and the Moon retains a record that extends over 4 billion years (perhaps earlier if there was no terminal cataclysm). On the Moon, however, impacts pulverize or melt most impactor material while repeatedly mixing the regolith and megaregolith. Very small amounts survive as mineral grains or rock fragments. Nevertheless, some materials, such as zircons, are particularly refractory and may be preserved in lunar regolith. Nonlunar zircons would be geochemically distinctive and, if found, they would provide information about timing, composition, and conditions on their parent body, possibly including the habitability of early Earth. Existing lunar samples might be used to evaluate whether such materials exist in a detectable amount and, if promising, methods could be developed for prospecting likely environments on the Moon and processing large amounts of lunar material to search for unusual components foreign to the Moon that could be returned to Earth for advanced forms of analysis.

The Moon may also constitute a testbed for astrobiological studies. The isolation of the Moon from Earth can be biologically quite complete, which could make the Moon an appropriate place to study biota in a fully sterile and sterilizing environment. This could include developing a lunar facility for assessing the potential biological impact of martian materials to Earth's environment. The exceedingly low organic levels in lunar regolith also may allow the Moon to be a useful testbed for very sensitive life-detection technologies.

EARTH SCIENCE FROM OR NEAR THE MOON

Remote Sensing of Earth From or Near the Moon

Earth observations from the Moon offer the another opportunity to acquire unique data to look at Earth in a bulk thermodynamic sense, particularly as an open system exchanging radiative energy with the Sun and space, in a way never done before—"Earth as a whole planet" (as illustrated in Figure 6.2). This is a fundamental scientific goal with very appealing prospects for climate and Earth sciences in general. Climate research requires stable,

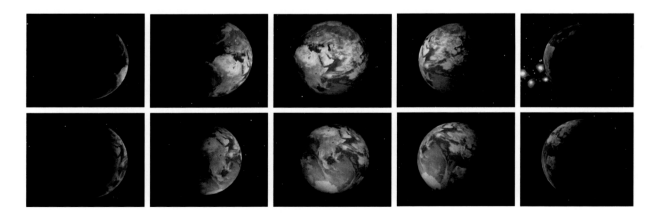

FIGURE 6.2 Simulated images of various phases of Earth as seen from the Moon over a period of days in both summer (top) and winter (bottom). SOURCE: Graphics created by Starry Night® and are used with permission from Imaginova Corporation. Starry Night is a registered trademark of Imaginova Corporation.

accurate, long-term observations made with adequate spatial and temporal resolution in a synoptic context. From lunar vantage points, it would be possible to sample the outgoing energy from virtually an entire hemisphere of Earth at once with high temporal and spatial resolution. At present this is only partially possible by combining data from LEO and geostationary orbit (GEO) satellites into an asynoptic composite of hundreds of thousands of pixels—rather like assembling an enormous jigsaw puzzle. Because of the integral view of the planet's hemispheres continuously in the infrared and periodically at visible wavelengths, the observations will simultaneously overlap the observations of every LEO and GEO satellite in existence, making possible a unique synergy with great potential benefits for Earth sciences. One of the major potential benefits is the enabling of calibration sharing with all satellites, thus allowing the integration of all Earth-observing satellites in a single, integrated Earth-observing system that would provide self-consistent enriched data sets for Earth sciences. Such synergism would undoubtedly represent a major advance in Earth sciences and a greatly enhanced return for the nation's investment in space and in particular for its investment in the exploration of the Moon.

Spectral observations from the Moon would allow for the first time continuous hemispheric synoptic retrievals of fundamental climate parameters, as well as data on cloud cover, aerosols, water vapor, and other atmospheric constituents, thus enabling unique observations of the diurnal cycle and its effects on the atmosphere. Note that observations in the IR are possible from all lunar orbital positions, while observations at visible wavelengths will be somewhat restricted.

Examples of possible instruments for observing Earth from the Moon are as follows:

• Telescopes as a front end for spectroscopes to provide high spatial resolution imaging,
• High-spectral-resolution spectrometers and/or interferometers covering the ultraviolet-visible-infrared spectrum, and
• Reference radiometers.

The lunar view is nicely balanced in that it shows Earth rotating as well as going through its day-night phase alternation every 27 Earth days. Thus, both visual near-infrared observations that show solar reflected light and mid-IR observations that show Earth temperature could be obtained and would give a complete view to reveal changes such as El Niño–La Niña alternations and longer-term changes, and to reveal changes in the albedo of clouds with time.

Observations of Earth at relatively fine scale are possible with existing instrumentation. For example, a modest 2048 × 2048 pixel coverage of Earth has a pixel width of about 6.2 km. Yet the angular resolution required for this is possible with a 4 cm aperture at visible wavelengths. Even a resolution 10 times coarser would be adequate to put ~75 pixels across the United States and to watch weather patterns moving across Earth. Such observations would usefully allow cross calibration of the various geosynchronous orbiting weather satellite observations at 0.1 percent precision and so permit issues of the phase functions for observations to be resolved without recourse to theory.

In Situ Observations of the Variable Sun

Lunar exploration will provide opportunities to make use of the Moon itself to help develop a unified understanding of the radiative variability of the Sun (e.g., the constancy of the "solar constant") on timescales of centuries. Drilling into the Moon's regolith, if extended to a depth of 10 meters, will enable a measurement of the borehole temperature profile. The intent of the borehole heat flux experiment is to derive the history of the solar constant from the present-day temperature profile in the borehole and to gain knowledge of the thermal diffusivity of the regolith. The physics here involves the analysis of the transport of heat from the lunar surface down into the regolith. The effect of the constant heat flow from the lunar interior is removed by looking at deviations from the linear profile. This experiment was already attempted by Apollo 15 and Apollo 17, which succeeded in measuring the diffusivity but could only measure temperature to a depth of 2.7 m. Interpretation of a lunar borehole thermal profile is straightforward, providing a measure of variations in the solar constant extending back to the time of Galileo's original 1610 mapping of sunspot areas, across the Maunder Minimum of sunspot numbers and the "Little Ice Age." Such an experiment would resolve the conflict among more indirect measures of the history of the solar constant, some indicating a significant brightening of the Sun since Galileo's time and some claiming no brightening.

The information about the variation of the solar constant, derivable from the Moon's regolith temperature profile, has important implications for the interpretation of Earth's currently observed global warming and is a prerequisite to any estimate of humanity's contribution to global warming.

HELIOPHYSICS OBSERVATIONS FROM OR NEAR THE MOON

Imaging of Radio Emissions from Solar Coronal Mass Ejections and Solar Flares

Radio observations of solar eruptions have played an important role in understanding the Sun, but the terrestrial ionosphere blocks all radio frequencies below 10 MHz to 20 MHz, which cover virtually all radio emissions originating above 1 to 2 solar radii from the Sun's surface.

Multi-site observations of low-frequency radio signals from the Moon, where there is no blocking ionosphere, would allow imaging of the sites of particle acceleration in the extended solar corona. In this region, the primary radio sources are fast (2 keV to 20 keV) electrons from solar flares and suprathermal electrons (~100 eV) accelerated by shocks. The associated radio emissions are called Type III bursts and Type II bursts, respectively. Both sources produce a plasma instability, which leads to amplification of electrostatic waves, some of which are then converted to electromagnetic (radio) waves. The process takes place at the characteristic frequency of the plasma—the electron plasma frequency—thus, the frequency of the radio emission indicates directly the electron density of the source, and imaging the radio source would map the extent of the acceleration region.

To make such images at low frequencies would require a synthetic aperture that is large compared with the wavelengths of interest. An angular resolution of 1 degree at 1 MHz requires a minimum diameter of 15 km. The Moon offers a large, stable surface on which to build a large, capable low-frequency radio array for the purpose of imaging solar sources at wavelengths that cannot be observed from the ground, an array that is well beyond the current state of the art for antennas in space. Figure 6.1 illustrates an antenna concept, operating at somewhat higher frequencies for radioastronomy purposes, being deployed on the Moon.

The design for the first (test) array would be based on the designs of Earth-based arrays (working at higher frequencies) currently in development. The deployment of the initial test elements could be done by astronauts or robotic deployers or both. Subsequent deployment of additional elements of the array would be carried out over time, permitting one to maximize lessons learned from implementing each phase.

The major components of the observatory are the antenna array and electric connections, the radio receivers, the central processing unit, the high-gain antenna unit, and the power unit. For a test array with 16 to 32 antennas, the expected total mass for a lunar observatory would be significantly less than 400 kg, and the power requirement would be less than 500 W.

Imaging of Earth's Ionosphere and Magnetosphere

Were it visible to the naked eye, Earth's magnetosphere would be the most spectacular object in the lunar sky. Subtending about 20 degrees at Full Earth and gradually expanding to envelop the entire lunar sky near New Earth, the magnetosphere would be an awe-inspiring phenomenon. Elements of the magnetosphere can, however, be imaged from the Moon through the use of modern instrumentation on timescales and at spatial resolutions of great interest to magnetospheric plasma physicists and specialists in space weather analysis and forecasting.

A unique advantage of the lunar perspective would be the ability to stare at the space neighborhood of Earth, providing global long-term images of the time development of extended magnetospheric phenomena such as the plasmasphere and ring current. The time period for significant evolution of these phenomena is a few hours, a time interval over which the perspective of an Earth-orbiting satellite changes significantly, while observations from the Moon do not vary appreciably in perspective.

Since the launch of the Imager for Magnetopause-to-Aurora Global Exploration (IMAGE) satellite in 2000, images of Earth's magnetosphere and ionosphere have become powerful tools in the investigation of the global dynamics of geospace. While high-latitude phenomena, such as the aurora, cannot be imaged all of the time effectively from the Moon, many other important magnetospheric and ionospheric phenomena at lower latitudes can be imaged for sufficiently long times to be of great interest. The lunar perspective would provide unique information in the area of space weather diagnostics.

The basic tools for geospace imaging are ultraviolet radiation and energetic neutral atom (ENA) cameras. With UV, emissions from specific ion species can be used to image large regions of the magnetosphere and ionosphere and to study the dynamics of these regions as the ions interact with terrestrial electric and magnetic fields.

ENA imaging is possible because Earth's hydrogen exosphere extends throughout large portions of the inner magnetosphere, as was demonstrated dramatically by the Apollo images obtained at the Lyman alpha wavelength. All ion populations in the magnetosphere undergo charge-exchange interactions with the exospheric atoms with the result that a few percent of the ion population is converted to energetic neutral atoms. Since these atoms are not affected by Earth's magnetic field, they can be imaged just as photons are imaged. Images formed by energetic neutral atoms provide global views of magnetospheric ion dynamics sorted by energy, velocity, and species.

Ultraviolet imaging from the Moon can reveal the dynamics of the plasmasphere, which is the upward extension of the ionosphere. The plasmasphere consists principally (about 85 percent) of protons, which cannot be imaged, but the remaining fraction is mostly He^+ ions, which can be imaged through resonantly scattered sunlight at a wavelength of 30.4 nanometers. Plasmasphere images are crucial space weather diagnostics, because during magnetic storms, giant plumes of plasma grow out of the plasmasphere as the plasmasphere itself shrinks. These plumes map along magnetic field lines into the ionosphere, producing enhancements of total electron content, which have been shown to seriously disrupt communications as well as Global Positioning System navigation signals. Examples of plasmasphere images taken from a low-latitude perspective are shown in Figure 6.3.

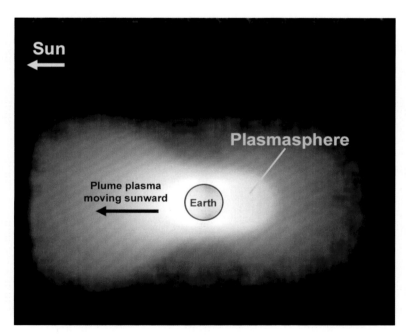

FIGURE 6.3 Simulated view of the He+ portion of Earth's plasmasphere, as it would appear to a proposed second-generation Imager for Magnetopause-to-Aurora Global Exploration (IMAGE)-Extreme Ultraviolet instrument camera system placed on the lunar surface. SOURCE: Courtesy of Dennis Gallagher, NASA Marshall Space Flight Center.

Similarly, the low-latitude ionosphere can be imaged from the Moon by detecting the UV emissions of ionospheric ions. Emissions from O^+ ions have been used to track the formation and evolution of density troughs associated with the ionospheric spread-F phenomenon. The motions of the density troughs, as shown in Figure 6.4, can be used to measure the evolution of large-scale ionospheric electric fields.

Energetic neutral atom imaging was shown to be effective during the IMAGE mission over a very large energy range, from 10 eV to 500 keV. One of the most important imaging targets is the ring current, which builds up in timescales of hours during magnetic storms and is a key ingredient in the dynamics of Earth's magnetosphere. An example of a ring current image taken from a low-latitude perspective is shown in Figure 6.5.

In summary, imaging from the Moon will provide a new perspective on magnetospheric and ionospheric dynamics because of the possibility of making observations from a vantage point providing a global view and for time spans substantially longer than the timescales of interest for these very dynamic phenomena.

The instrumentation for lunar-based geospace imaging would be based on the flight-proven IMAGE instruments with correspondingly larger apertures to allow for the approximately seven times larger distance. Multispectral UV and ENA imagers would require relatively modest resources, in the range of 50 kg to 100 kg and 30 W to 50 W each. Deployment could be done either robotically or by astronauts.

Finding 5: The Moon may provide a unique location for observation and study of Earth, near-Earth space, and the universe.

The Moon is a platform that can potentially be used to make observations of Earth (Earth science) and to collect data for heliophysics, astrophysics, and astrobiology. Locations on the Moon provide both advantages and disadvantages. There are substantial uncertainties in the benefits and the costs of using the Moon as an observation platform as compared with alternate locations in space. The present committee did not have the required span

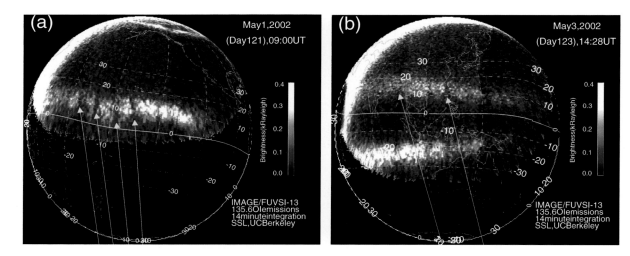

FIGURE 6.4 Oxygen airglow images for (a) May 1, 2002, and (b) May 3, 2002, showing nightside equatorial plasma bubbles whose drift motion can be used to determine ionospheric electric fields. (The arrows identify "bubbles" in the ionosphere, which are used to determine ambient plasma drift speeds in the second part of the figure, not shown here.) Data from Imager for Magnetopause-to-Aurora Global Exploration (IMAGE)-far-ultraviolet instrument. SOURCE: Courtesy of T.J. Immel, H.U. Frey, S.B. Mende, and E. Sagawa, Global observations of the zonal drift speed of equatorial ionospheric plasma bubbles, *Ann. Geophys.* 22:3099-3107, 2004.

FIGURE 6.5 Energetic neutral atom (ENA) image at ~20 keV to 50 keV from an equatorial perspective. Data from Imager for Magnetopause-to-Aurora Global Exploration (IMAGE)-high-energy neutral-atom imager instrument. SOURCE: Courtesy of J. Burch, Southwest Research Institute, D. Mitchell, and P. Brandt with support from NASA under contract NAS5-96020.

and depth of expertise to perform a thorough evaluation of the many issues that need examination. A thorough study is required.

Recommendation 5: The committee recommends that NASA consult scientific experts to evaluate the suitability of the Moon as an observational site for studies of Earth, heliophysics, astronomy, astrophysics, and astrobiology. Such a study should refer to prior NRC decadal surveys and their established priorities.

7

Concepts Related to the Implementation of Science

The committee identified several related concepts (numbered 1R through 4R) pertaining to the optimal implementation of science in the Vision for Space Exploration (VSE). This effort was driven in large part by the stark realization that more than 30 years have passed since Apollo and that the nature of the VSE itself warrants a major reconsideration of the basic approach to conducting lunar science. In more than 30 years, robotic capability has increased dramatically, analytical instrumentation has advanced remarkably, and the very understanding of how to explore has evolved as scientists have learned about planetary formation and evolution. The VSE offers new opportunities: researchers are no longer limited to short-duration lunar stays of 2 or 3 days and "emplacement science"; scientists on the Moon can operate as scientists, doing analytical work and deciphering sample/source relationships; site revisit with follow-up science is possible (e.g., an outpost); robotic-capable equipment can be used between missions; geophysical equipment can be used in survey modes; time-consuming deep drilling is possible; lunar samples can be "high-graded" for return to Earth. These are but examples of how different the VSE exploration should be as compared with that of the Apollo program. Nurturing a new approach to lunar exploration must be fostered early in the program if we are to reap the potential of the VSE is to be reaped.

CONCEPTS RELATED TO OPTIMAL IMPLEMENTATION OF SCIENCE IN THE VISION FOR SPACE EXPLORATION

Concept 1R: Managing Science in a Program of Human Exploration

Successful implementation of science in a program of human exploration starts with and is highly dependent on a cooperative relationship among all of the involved communities. To acquire lessons learned from past experience, the Space Studies Board's Committee on Human Exploration (CHEX) conducted a study of science prerequisites, science opportunities, and science management in the human exploration of space.[1] For science management, CHEX studied the Apollo, Skylab, Apollo-Soyuz, and Shuttle/Spacelab programs to determine what organizational relationships, roles, and responsibilities contributed to superior science outcomes.

---

[1]These National Research Council reports from SSB's CHEX are *Scientific Prerequisites for the Human Exploration of Space* (1993), *Scientific Opportunities in the Human Exploration of Space* (1994), and *Scientific Management in the Human Exploration of Space* (1997), published by the National Academy Press, Washington, D.C.

CHEX found that human exploration offers a unique opportunity for science accomplishment and as such should be viewed as part and parcel of an integrated human exploration-science program. That committee developed three broad management principles, which, if implemented, would improve the probability of a successful synergy between science and human exploration:

1. *Integrated Science Program*—The scientific study of specific planetary bodies, such as the Moon and Mars, should be treated as an integral part of an overall solar system science program and not separated out simply because there may be concurrent interest in human exploration of those bodies. Thus, there should be a single NASA headquarters office responsible for conducting the scientific aspects of solar system exploration.

2. *Clear Program Goals and Priorities*—A program of human spaceflight will have political, engineering, and technological goals in addition to its scientific goals. To avoid confusion and misunderstandings, the objectives of each individual component project or mission that integrates space science and human spaceflight should be clearly specified and prioritized.[2]

3. *Joint Spaceflight/Science Program Office*—The offices responsible for human spaceflight and space science should jointly establish and staff a program office to collaboratively implement the scientific component of human exploration. As a model, that office should have responsibilities, functions, and reporting relationships similar to those that supported science in the Apollo, Skylab, and Apollo-Soyuz Test Project (ASTP) missions.

Consistent with the principles enunciated above, CHEX found a definitive correlation between successful science accomplishment and organizational roles and responsibilities. In particular, the quality of science was enhanced when the science office (NASA's Science Mission Directorate [SMD] now) controlled the process of establishing science priorities, competitively selecting the science and participating scientists and ensuring proper attention to the end-to-end cycle ending in data analysis and the publication of results. This process extends to the selection of science contributions by international partners; competitive merit should prevail.

Finding 1R: The successful integration of science into programs of human exploration has historically been a challenge. It remains so for the VSE. Prior Space Studies Board reports by the Committee on Human Exploration (CHEX) examined how the different management approaches led to different degrees of success. CHEX developed principles for optimizing the integration of science into human exploration and recommended implementation of these principles in future programs.[3] This committee adopts in Recommendation 1R the CHEX findings in a form appropriate for the early phase of VSE.

Recommendation 1R: NASA should increase the potential to successfully accomplish science in the VSE by (1) developing an integrated human/robotic science strategy,[4] (2) clearly stating where science fits in the Exploration Systems Mission Directorate's (ESMD's) goals and priorities, and (3) establishing a science office embedded in the ESMD to plan and implement science in the VSE. Following the Apollo model, such an office should report jointly to the Science Mission Directorate and the ESMD, with the science office controlling the proven end-to-end science process.

Concept 2R: Developing Lunar Mission Plans and Operations

Apollo experience demonstrated both the complexity of planning lunar surface and orbital operations and the benefits of so doing. The challenge today is, if anything, greater, in that the more than 30 year hiatus since Apollo has seen a remarkable evolution of planetary exploration strategy and capability. One cannot start too

---

[2]See especially pp. 2-3 in the section "The Role of Science" in the 1993 report *Scientific Prerequisites for the Human Exploration of Space*; pp. 6-7 in the 1994 report *Scientific Opportunities in the Human Exploration of Space*; and pp. 17-29 in Chapter 3, "Science Enabled by Human Exploration," in the 1994 report.

[3]See p. 128 of the third report in a series by the Committee on Human Exploration: National Research Council, *Science Management in the Human Exploration of Space*, National Academy Press, Washington, D.C., 1997.

[4]This CHEX Recommendation 1 refers to the development of science goals, strategy, priorities, and process methodology; CHEX Recommendation 3 and this committee's Recommendation 1R refers strictly to the implementation of science in a program of human exploration.

early: detailed Apollo planning started in the early 1960s, well before Apollo 11 landed on the Moon in 1969, and included robotic precursors, determination of science objectives, astronaut selection, selection of landing sites, astronaut science training, science team selection, traverse planning, sampling strategies, geophysical station development (the Apollo Lunar Surface Experiments Package [ALSEPs]), full-up mission simulations, and data-analysis preparations. A similar range of preparation is essential for implementation of the Vision for Space Exploration. The merit of including scientists in the astronaut corps has long been recognized and the benefits demonstrated on Apollo 17, Skylab, the Space Shuttle, Spacelab, and the International Space Station. Scientist-astronauts should be among all lunar mission crews.

Much of what was learned by hard experience on Apollo can now be efficiently incorporated into VSE planning. Obviously much of the detailed planning will have to, and should, await a date closer to mission implementation; however, many aspects can be initiated now with relatively low investment and high return. Two areas stand out as needing early planning and the involvement of the science community: (1) site selection and the related issue of sortie missions and/or a lunar outpost, and (2) surface mission planning and related requirements for mobility and the use of robotics.

1. *Landing site selection.* NASA's plans to return humans to the Moon necessarily involve the selection of surface exploration sites. The site selection process becomes a matter of which sites best satisfy Exploration Systems Mission Directorate (ESMD) goals and priorities. Among the considerations are science, the buildup of an outpost or base, preparation for Mars exploration, the development of in situ resource utilization (ISRU), and commercial potential. Then will come the obvious overlay of budget and engineering limitations, site access, logistics, and safety. Successful accomplishment of many of the science goals elucidated in this report depends critically on getting to specific lunar landing sites. In contrast to site selection for the Apollo program, VSE site selection has to satisfy multiple goals of which science is but one, albeit an important one. The challenge becomes one of optimizing site selection to accomplish multiple goals. The committee notes that there are site selection considerations that are independent of human exploration sites: for example, robotic sample return from sites that may not be visited by humans and/or the global emplacement of geophysical networks. These science activities recognize that the VSE is advertised as not solely for human missions but that it is to involve an ongoing mix of human and robotic missions.

A dichotomy already exists regarding lunar landing sites. On the one hand, the sortie mode is preferred by most lunar scientists, who consider it necessary to visit many diverse lunar sites both for geologic studies and for instrument emplacement. The ESMD, on the other hand, has made a preliminary determination that a singular (tentatively polar region) outpost site best serves its higher-priority goals of "habitation" and "prepare for exploration." Although a sortie capability is currently stated as continuing to be available, the cost of such a capability would be billed to the Science Mission Directorate (at about $2 billion). That cost must be traded off against accomplishing the science goals robotically and against competing nonlunar space science.

The attributes of the lunar outpost concept for purposes of scientific investigation deserve joint ESMD/SMD study. The potential advantages are increased time for detailed geologic study; deep drilling, core retrieval, and downhole instrument emplacement; geophysical instrument emplacements; traverse surveys; returned-sample selection ("high-grading"); follow-up on results obtained on earlier outpost missions; and utilization of logistics previously emplaced. An outpost would warrant a greater investment in terms of reusable resources—for example, a multimission rover with resuppliable onboard life support and sophisticated analytical instrumentation. Such a rover could be used in automated mode between outpost visits. Many of these attributes were being considered in the 1960s as follow-ons to the initial Apollo missions but were, obviously, never executed when missions after Apollo 17 were canceled. It is important to note that the precise location of an outpost site will determine the scientific return. It thus behooves the ESMD to incorporate scientific site criteria among its overall criteria.

2. *Surface and orbital mission operational planning.* Apollo experience demonstrated that the most valuable resource on the Moon is time. There is inevitably more to be done than time allows. For example, astronauts were constantly under pressure to "move on to the next station." Many opportunities to examine discoveries in more detail were missed. Things that went wrong (e.g., a stuck drill or an instrument failing to operate) took time away from meeting the time lines. It is necessary to devise methods to conserve astronaut time, doing robotically

those things that do not take greatest advantage of the human capability to observe, to make decisions, and to use manual dexterity to advantage. NASA should undertake a reexamination of Apollo missions with an eye toward identifying time-saving opportunities. It should examine roles for the much-improved robotics capability available today, including consideration of robotic assistants.

Astronauts should be employed as an inherent on-site element in early surface science activities, when the unequaled human characteristics of adaptability, quick reaction, and discerning observation offer the best potential for achieving a science objective. In all considerations of on-the-surface human versus telepresence versus autonomous implementation, trade-offs will compare cost, capability, safety, probability of task success, and coincident activity. Mission planning and management tools for this specific purpose should be developed.

The typical Apollo-type sortie mission of 2 to 3 days is too short to accomplish the level of scientific investigation now merited by our improved understanding of lunar science. If, however, sortie missions are a selected mode for early exploration missions (currently, sortie capability of up to 7 days is under consideration), then planning is needed to increase their efficiency. A possible mode is to precede the human flight with robotic rover precursors: the rovers, possibly similar to the Mars Exploration Rovers (MERs) on Mars, would conduct reconnaissance and identify high-priority traverse locations for astronaut investigation. The Lunar Roving Vehicles on Apollo missions 15 to 17 demonstrated the benefit of mobility; their range was limited primarily by safety and life-support supply considerations. Analysis of the extent to which such constraints can be relieved on future missions is necessary. The desirability of increasing mobility range beyond the canonical walk-back distance (about 10 km) is supported by the general observation that the lunar geologic variety occurs on the scale of tens to hundreds of kilometers. Given the paramount consideration of safety and limited time, telerobotic operations during a human mission might add immeasurably to mission efficiency. Surface rovers capable of being either teleoperated or crewed, (so-called dual mode) should be developed and used on all surface missions. Outfitted with manipulators and observational/analytical instrumentation, these rovers would become telerobotic explorers when astronauts are not present. And with a capability for long-range travel, they potentially could be redeployed from one sortie science site to another, thus saving mission systems duplication. Such site-to-site traverses offer a great potential for geophysical profiling and geological observations of opportunity in a true discovery mode.

Finding 2R: Great strides and major advances in robotics, space and information technology, and exploration techniques have been made since Apollo. These changes are accompanied by a greatly evolved understanding of and approach to planetary science and improvements in use of remote sensing and field and laboratory sample analyses. Critical to achieving high science return in Apollo was the selection of the lunar landing sites and the involvement of the science community in that process. Similarly, the scientific community's involvement in detailed mission planning and implementation resulted in efficient and productive surface traverses and instrument deployments.

> **Recommendation 2R:** The development of a comprehensive process for lunar landing site selection that addresses the science goals of Table 5.1 in this report should be started by a science definition team. The choice of specific sites should be permitted to evolve as the understanding of lunar science progresses through the refinement of science goals and the analysis of existing and newly acquired data. Final selection should be done with the full input of the science community in order to optimize the science return while meeting engineering and safety constraints. Similarly, science mission planning should proceed with the broad involvement of the science and engineering communities. The science should be designed and implemented as an integrated human/robotic program employing the best each has to offer. Extensive crew training and mission simulation should be initiated early to help devise optimum exploration strategies.

Concept 3R: Identifying and Developing Advanced Technology and Instrumentation

The preceding section on Concept 2R discusses a number of operational concepts, equipment, instruments, and analytical tools that will enhance VSE lunar science. The eventual incorporation of many of these, and others not yet determined, into missions is dependent on the development of advanced technology and instrumentation.

Given the long lead time associated with advancing technology, it is imperative to start soon to fully identify, prioritize, and fund those enabling technologies. When such technologies are successfully developed through NASA strategic investments, they should have a clear path into flight development to meet either SMD or ESMD goals and requirements. Providing well-defined paths into flight development will make the best use of and leverage NASA technology development funds as well as provide an incentive for other technology providers (i.e., industrial, academic, governmental) to take their own risk of developing new technologies. To give a sense of the variety and magnitude of the needed developments, some of the most critical technologies are enumerated below.

1. *Advanced technology.* The combination of surface mobility and equipment and sample manipulation is a key requirement for conducting science on the Moon, during both precursor and sortie missions. For autonomous/semi-autonomous robotic operations on the lunar surface, large gaps exist in current robotic capabilities. These gaps should be addressed to enable the achievement of science goals. The challenges include development of the following capabilities:

- Long-distance traversing, navigation, and access:
  —Rovers (possibly dual-mode with crewed vehicles) with multiyear life, and day/night and permanent-shadow operational capability;
  —Rover capability to access and maneuver on all lunar terrain types, including disturbed lunar soil, steep slopes, craters, and basins; to traverse long distances; to carry/deploy payloads; and to cache samples for later retrieval;
  —Navigation in shadowed regions, such as those found in the polar craters;
  —Enhanced visualization and assessment of the environment for human supervision/telepresence; and
  —Communications infrastructure for full-time teleoperation of surface mobility.
- Instrument placement and manipulation:
  —Dexterous placement of science instruments;
  —Interchange of end-effectors needed to achieve contact measurements;
  —Robust acquisition, manipulation, and analysis of multiple science samples and transport to a location of interest (e.g., to a sample cache site);
  —Manipulation of sensors for active experimentation such as seismic and electromagnetic sounding;
  —Subsurface data collection: drilling, core retrieval, and downhole emplacement of instruments;
  —Self-contained experiment packages (a la Shuttle Hitch-Hikers and Get Away Specials) requiring minimal crew resources for deployment; and
  —Geophysical probe delivery systems enabling orbital deployment at globally distributed locations.

2. *Instrumentation.* Some types of in situ and laboratory measurement technology have not yet achieved their potential to contribute to the accomplishment of scientific goals. For use on both robotic and human missions and for returned samples, development of the following instrumentation capabilities is needed:

- In situ determination of the radiometric age of a crystalline igneous or impact melt;
- In situ measurement of cosmic-ray exposure ages;
- In situ routine microanalysis (major elements, mineralogy) and imaging at the 10 micron scale, for fields of view of a few millimeters;
- In situ measurement of minor and trace elements for gram-sized samples;
- High-resolution remote sensing, at the scale of tens of centimeters, to allow an assessment of local geology and precise targeting of samples, and to inform the crews and ground support of the types and distributions of materials present at the site;
- Upgrading of analytical instrumentation for sample analyses in the curatorial facilities and in principal investigators' laboratories (see also Concept 4R); and
- Geophysical instrumentation capable of being deployed from orbit and surviving high-g impacts.

Finding 3R: The opportunity provided by the VSE to accomplish science, lunar and otherwise, is highly dependent for success on modernizing the technology and instrumentation available. The virtual lack of a lunar science program and no human exploration over the past 30 years have resulted in a severe lack of qualified instrumentation suitable for the lunar environment. Without such instrumentation, the full and promising potential of the VSE will not be realized.

> **Recommendation 3R:** NASA, with the intimate involvement of the science community, should immediately initiate a program to develop and upgrade technology and instrumentation that will enable the full potential of the VSE. Such a program must identify the full set of requirements as related to achieving priority science objectives and prioritize these requirements in the context of programmatic constraints. In addition, NASA should capitalize on its technology development investments by providing a clear path into flight development.

Concept 4R: Updating Lunar Sample Collection Techniques and Curation Capabilities

The NASA system of sample documentation and curation established for the Apollo program has been remarkably successful in protecting samples from contamination, providing ample materials for scientific investigations, preserving materials for later studies, and maintaining configuration control of the collection so that analyses of subsamples can be reliably related to the samples from which they were derived or to other subsamples. It has also been the model on which the means of documentation, preservation, and subsampling of Antarctic meteorites, cosmic dust, and Stardust and Genesis samples were based. The Apollo lunar samples have always been treated as if they were the last samples that might be retrieved from the Moon. The Vision for Space Exploration offers the possibility of collecting many more samples from the Moon, from a much wider variety of locations. This potentially includes the collection of samples from much more extreme environments than was possible during Apollo: examples are cryogenic samples from shadowed polar craters, samples preserved in lunar vacuum conditions, or electrostatically levitated dust. There is also the potential to return additional samples of human-made materials exposed to the lunar environment, which will be of scientific interest (for astrobiology and planetary protection) as well as engineering interest. It is important to re-evaluate the curatorial functions for lunar samples in light of the new opportunities and capabilities of the Vision for Space Exploration. Historically NASA has successfully involved the broad science community—for example, the Curation and Analysis Planning Team for Extraterrestrial Materials (CAPTEM) to assist it in such evaluations and assessments.

It is well to remember the principal objectives of proper extraterrestrial sample preservation: (1) to preserve information on the relationship of the sample to its original environment on the planetary body; (2) to ensure that samples, once collected and returned to Earth, are both available to study and preserved for future scientific investigations, which may be more rigorous with regard to precision, scale, and degree of contamination; and (3) to protect samples from unfavorable contamination due to handling techniques or exposure to the terrestrial environment.

This preservation process must begin on the Moon, with the collection, documentation, and packaging of samples. Documentation of the original collection location must be accomplished by crewmembers on the surface, provided with adequate documentation tools. During Apollo, the orientation of some rocks proved important in some studies of the lunar radiation environment. It will be especially important to document any samples collected in situ, that is, from outcrops, rather than loose materials in the regolith. Several samples from the same outcrop might be collected to study its internal variations, which must be kept separate if their original relations (position, orientation) to the outcrop are to be preserved. It proved close to impossible to preserve Apollo samples in lunar vacuum conditions (although there were few, if any, vacuum chambers on Earth that could be used to study samples maintained at lunar vacuum), because of problems of dust on container seals. Documentation of the deployment and functioning of special sample collection and containment devices, which will undoubtedly be developed for particular samples (such as cryogenic samples from the poles), and provision for transferring these samples to Earth, will be needed. The consideration of these requirements should be incorporated into the design of surface exploration tools provided to lunar crews.

Curation of lunar samples has always aimed to avoid cross-contamination of samples. A major manifestation of this requirement is that samples from more than one location on the Moon have been handled in separate work areas (typically nitrogen-filled stainless steel cabinets) or in work areas that are thoroughly cleaned before samples from another area are introduced. The challenge of working with samples from a much wider variety of areas will be significant, and the procedures and facilities for doing that should be re-evaluated. For example, working with a large number of small samples from a diversity of areas might require smaller, more flexible, work areas.

Many of the large samples collected by Apollo proved to be needlessly redundant scientifically. When most analyses can be done on milligram to gram quantities of samples, 10 kg rock samples are likely to be too large. This may not be true of rock samples for which important internal relationships (e.g., contacts between matrix and clasts in complex breccias) require study. Large regolith samples (perhaps greater than 200 g) can prove to be cumbersome to handle and preserve, when only a few grams of material are designated for study. These types of considerations should be included in designing crew surface procedures, sampling tools, and containers.

The ability to return to a field area or to a central outpost will open some opportunities not available to the Apollo program. In particular, capability can be provided for storing larger pieces of rock on the Moon in a protected environment, in case there is a future call for study of those samples when something unusual is discovered during scientific investigations. As mentioned elsewhere, analytical capability on the Moon could be used to select portions of samples, such as fragments from a sieved regolith sample. In these cases, the residual materials, which may have future scientific value, should be isolated and preserved in some manner. The characteristics of a lunar curatorial facility and associated hardware and procedures should receive study.

Finding 4R: The NASA curatorial facilities and staff have provided an exemplary capability since the Apollo program to take advantage of the scientific information inherent in extraterrestrial samples. The VSE has the potential to add significant demands on the curatorial facilities. The existing facilities and techniques are not sufficient to accommodate that demand and the new requirements that will ensue. Similarly, there is a need for new approaches to the acquisition of samples on lunar missions.

Recommendation 4R: NASA should conduct a thorough review of all aspects of sample curation, taking into account the differences between a lunar outpost-based program and the sortie approach taken by the Apollo missions. This review should start with a consideration of documentation, collection, and preservation procedures on the Moon and continue to a consideration of the facilities requirements for maintaining and analyzing the samples on Earth. NASA should enlist a broad group of scientists familiar with curatorial capabilities and the needs of lunar science, such as the Curation and Analysis Planning Team for Extraterrestrial Materials (CAPTEM), to assist it with the review.

8

Concluding Remarks

It is the unanimous consensus of the Committee on the Scientific Context for Exploration of the Moon that the Moon offers profound scientific value. The infrastructure provided by sustained human presence can enable remarkable science opportunities if those opportunities are evaluated and designed into the effort from the outset. While the expense of human exploration cannot likely be justified on the basis of science alone, the committee emphasizes that careful attention to science opportunity is very much in the interest of a stable and sustainable lunar program. In the opinion of the committee, a vigorous near-term robotic exploration program providing global access is central to the next phase of scientific exploration of the Moon and is necessary both to prepare for the efficient utilization of human presence and to maintain scientific momentum as this major national program moves forward.

Principal Finding: Lunar activities apply to broad scientific and exploration concerns.

Lunar science as described in this report has much broader implications than simply studying the Moon. For example, a better determination of the lunar impact flux during early solar system history would have profound implications for comprehending the evolution of the solar system, early Earth, and the origin and early evolution of life. A better understanding of the lunar interior would bear on models of planetary formation in general and on the origin of the Earth-Moon system in particular. And exploring the possibly ice-rich lunar poles could reveal important information about the history and distribution of solar system volatiles. Furthermore, although some of the committee's objectives are focused on lunar-specific questions, one of the basic principles of comparative planetology is that each world studied enables researchers to better understand other worlds, including our own. Improving our understanding of such processes as cratering and volcanism on the Moon will provide valuable points of comparison for these processes on the other terrestrial planets.

Bibliography

The Committee on the Scientific Context for Exploration of the Moon consulted related National Research Council reports issued by the Space Studies Board (SSB), with other boards as indicated. Each report was published by the National Academy Press (after mid-2002, The National Academies Press), Washington, D.C., in the year indicated.

Assessment of Options for Extending the Life of the Hubble Space Telescope: Final Report, SSB and Aeronautics and Space Engineering Board (ASEB) (2005)
Astronomy and Astrophysics in the New Millennium, SSB and Board on Physics and Astronomy (2001)
The Astrophysical Context of Life, SSB and Board on Life Sciences (2005)
An Integrated Strategy for the Planetary Sciences: 1995-2010 (1994)
Lessons Learned from the Clementine Mission (1997)
Lunar Exploration—Strategy for Research: 1969-1975 (1969)
New Frontiers in the Solar System: An Integrated Exploration Strategy (2003)
Priorities in Space Science Enabled by Nuclear Power and Propulsion, SSB and ASEB (2006)
Review of Goals and Plans for NASA's Space and Earth Sciences (2006)
The Role of Small Missions in Planetary and Lunar Exploration (1995)
Science in NASA's Vision for Space Exploration (2005)
Science Management in the Human Exploration of Space (1997)
Scientific Opportunities in the Human Exploration of Space (1994)
Scientific Prerequisites for the Human Exploration of Space (1993)
A Scientific Rationale for Mobility in Planetary Environments (1999)
The Search for Life's Origins: Progress and Future Directions in Planetary Biology and Chemical Evolution (1990)
Strategy for Exploration of the Inner Planets: 1977-1987 (1978)
Update to Strategy for Exploration of the Inner Planets (1990)

Appendixes

A

Statement of Task

BACKGROUND

The Moon is the first waypoint for human exploration in the Vision for Space Exploration. While not premised primarily on science goals, a well-planned and executed program of human exploration of the Moon and of the robotic missions that will precede and support it offers opportunities to accomplish important scientific investigations about the Moon and the solar system beyond.

NASA is aggressively defining and implementing the first missions in a series of robotic orbital and landed missions, the Lunar Precursor and Robotic Program (LPRP) of the Exploration Systems Mission Directorate (ESMD). The LPRP is intended to obtain essential supporting data for precursor robotic and human landings planned for 2018 and shortly thereafter. The first LPRP mission, the Lunar Reconnaissance Orbiter, is already in implementation and scheduled for a 2008 launch. A second mission, a lander, is in pre-formulation. The LPRP program office is currently developing an overall LPRP program architecture. Payloads for these forerunner robotic missions respond primarily to requirements for supporting robotic and future human landings, but may offer also opportunities to acquire scientifically valuable information as well. In order to realize this benefit from the LPRP series, NASA needs a comprehensive, well-validated, and prioritized set of scientific research objectives for the Moon.

Looking beyond the robotic precursor missions, science goals will also be needed to inform early decisions about system design and operations planning for human exploration of the Moon. In a near-term program of sortie-mode human landings with their capability for in situ instrument deployment and operation as well as informed sample return, the most immediate candidates for investigation are lunar science and the history of the solar system, including the history of the Sun. Design and planning for human exploration will need insight into the types of investigations that astronauts on the Moon might carry out as well as projections of necessary equipment and operations. The point of departure for this planning would be the Apollo program. However, NASA's current plans envisage spacecraft with superior capabilities and endurance to those of the Apollo program. For example, the new lunar landing vehicle may initially support a crew of four on the surface of the Moon for a week, compared to the Apollo landing vehicle's crew of two and surface stay time of 2-3 days.

For longer range human presence on the moon, the scope of science is potentially broader, possibly including emplacement or assembly and maintenance and operation of major equipment on the lunar surface. Expanded future presence could evolve from the near-term program by offering permanent, versus sortie, human presence and by a greatly increased landed mass on the Moon. Follow-on lunar-based science might include not only intensified lunar surface research, but also possibly observations of Earth and of the universe beyond the solar system. Eventually, analyses and trade studies on science efficacy and cost benefit will be required in order to understand the value of the Moon as a site for such undertakings.

STUDY SCOPE

The current study is intended to meet the near-term needs for science guidance for the lunar component of the Vision for Space Exploration. In the context of the above background, the *primary goals* of the study are to:

1. Identify a common set of prioritized basic science goals that could be addressed in the near-term via the LPRP program of orbital and landed robotic lunar missions (2008-2018) and in the early phase of human lunar exploration (nominally beginning in 2018); and

2. To the extent possible, suggest whether individual goals are most amenable to orbital measurements, in situ analysis or instrumentation, field observation or terrestrial analysis via documented sample return.

Secondary goals:

3. Goals and activities oriented toward ESMD requirements, for example, LPRP characterization of the lunar environment of value to human safety and in situ resource utilization (ISRU), should be analyzed, to the extent that these characterization requirements are provided by ESMD. These should be tabulated separately, but areas of overlap between basic science goals and these ESMD requirements should be noted as synergistic opportunities.

4. It is not intended that the current study address in depth more ambitious future opportunities that would entail assembly of large and complex research apparatus on the Moon. Examples are major lunar astronomical observatories or Earth observation systems that might follow systems currently in formulation or development. Implementation of such systems could become possible after the initial phases of human exploration. Science goals for astronomy and astrophysics are already provided by the NRC reports *Astronomy and Astrophysics in the New Millennium* (NRC, 2000) and *Connecting Quarks with the Cosmos: Eleven Science Questions for the New Century* (NRC, 2003). Earth system science and applications goals will be articulated in the new NRC decadal survey under development for this area and due for completion in late 2006. In these areas, the present study should limit itself to collecting and characterizing longer term possibilities, if any, that deserve feasibility and cost/benefit analysis in a future study.

The *science scope* of study goals 1 and 2 should encompass:

- The history of the Moon and of the Earth-Moon system;
- Implications for the origin and evolution of the solar system generally, including the Sun; and
- Implications of all of these for the origin and evolution of life on Earth and possibly elsewhere in the solar system.

Applied laboratory research in life sciences or materials or physical science in the lunar low-gravity environment oriented toward human Mars exploration requirements are not within scope of the task, but could be addressed in a future study.

Where appropriate, activities recommended for implementation within the lunar exploration component of the Vision for Space Exploration should be compared to other means of implementing the same scientific goals. There is a broad spectrum of science ideas being discussed for the lunar program at this time. The intent is that the committee focus on the strongest and most compelling ideas that come before it. There is a broad and expert community of planetary scientists with special interest in the Moon and lunar science. This community will make scientifically persuasive arguments that certain lines of inquiry can be uniquely well-conducted on the Moon. It is anticipated that the goal prioritization requested for this study will differentiate between science investigations that can only be done on the Moon, those that could potentially be competitively conducted on the Moon depending on analyses of cost and technical factors, and investigations for which current knowledge and forecasted capabilities lend little support for lunar implementation. It is essential that NASA adopt the very strongest science program possible for the Moon right from the outset because advocated weak science would be questioned and could jeopardize the entire lunar program.

While premised on a framework of essentially flat science budgets in the near term, the study may consider also the possibility of expanded budgets for lunar science in the post-2010 time frame, after shuttle retirement.

Because lunar exploration within the Vision for Space Exploration is envisioned as a broadly international undertaking, the study should attempt to factor in interests and perspectives of foreign investigators and/or agency officials by inclusion of some of these as panel members and/or as briefers, as appropriate.

DELIVERABLES AND SCHEDULE

It is anticipated that development of the study products will be undertaken via a two phase process consisting of (1) an initial review and integration of goals and priorities in existing NRC and other documentation and (2) a science community outreach program to validate, update, and extend the findings of this review to support planning for potential follow-on LPRP missions and astronaut missions during the sortie phase of human lunar exploration.

The final report of the study should contain the following primary elements:

1. A brief summary of the current status and key issues of scientific knowledge concerning the origin and evolution of the Moon and related issues in solar system evolution;

2. Basic science goals and priorities for research within scope of the study task that are contained in NRC decadal surveys relevant to lunar exploration, as expanded and extended.

Insofar as is possible, the final report should also contain:

3. A summary of the scientific measurements and LPRP activities necessary to support the safe return of humans to the Moon, to the extent that relevant requirements are provided to the NRC by ESMD; and

4. A high-level survey of possible future activities and infrastructure that could address science objectives lying outside the current study scope but deserving of feasibility and cost/benefit evaluation for potential implementation during a long-term human presence on the Moon.

An interim description of basic science goals and priorities (items 2 and 3 above) should be released for community discussion and preliminary NASA planning use, if possible by *August 31, 2006*. Delivery at this time would enable its use by NASA to inform finalization of the fiscal year 2008 budget proposal, and to support an exploration strategy drafting meeting being planned by ESMD for mid-September 2006 and a NASA Advisory Council Science Committee workshop planned for later in 2006.

A prepublication version of the final report should be delivered by *May 31, 2007*, in order to be of maximum value during formulation of the fiscal year 2009 budget proposal.

The present task may not extend beyond July 14, 2007. Delivery of the edited final report will be negotiated once the study is in progress and may be supported by a task issued under the follow-on contract.

B

Glossary, Acronyms, and Abbreviations

ALSEP—Apollo Lunar Surface Experiments Package

anorthosite—a type of rock made up mainly of plagioclase feldspar, which has been found in all the Apollo lunar samples and constitutes most of the light-colored crust on the Moon

Archaea—a recently recognized domain of prokaryotic life. Single-celled microorganisms that lack a nucleus, are morphologically similar to bacteria but not closely related, having features such as genetic transcription and translation that are different. They are found in very high temperature environments, and include methanogens and hyperthermophiles that may be similar to the first life on Earth

ASTP—Apollo-Soyuz Test Project, a joint U.S.–U.S.S.R. mission

ATP—Advanced Technology Program of the National Aeronautics and Space Administration (NASA)

bacteria—single-celled micro-organisms lacking a nucleus, morphologically similar to Archaea, but not closely related

breccia—a rock composed of angular fragments of rocks and minerals in a matrix

CAPTEM—Curation and Analysis Planning Team for Extraterrestrial Materials

Chang'e—a Chinese National Space Administration lunar orbiter

Chesapeake—an ancient subsurface impact crater located in the Chesapeake Bay

CHEX—National Research Council's Committee on Human Exploration

Chicxulub—an ancient subsurface impact crater located near Chicxulub, Mexico

chondrite—a type of stony meteorite containing chondrules, roughly spherical bodies containing pyroxene or olivine embedded in the matrix

chronostratigraphy—the branch of stratigraphy that studies the absolute age of rocks

cool early Earth—hypothesis that the surface of Earth cooled relatively quickly after the formation of the core and the Moon, such that oceans and conditions hospitable for life could exist by 4.3 Ga

Copernican—the lunar geologic period from about 1.1 Ga ago to the present

Cretaceous/Tertiary boundary—on Earth, the boundary between rocks of the Cretaceous and Tertiary periods, about 65.5 million years ago, around the time of a major extinction event

cumulates—igneous rocks formed by the accumulation of crystals from a magma

detrital zircons—zircon crystals found in erosional deposits. Some grains are as old as 4.4 Ga and are the oldest known samples of Earth

differentiated planetary body—a planetary body whose interior is formed of separate internal geologic units with distinct mineralogical characteristics, e.g., core, mantle, crust

Erastosthenian—the lunar geologic period from 1.1 Ga to 3.2 Ga ago

ESMD—NASA's Exploration Systems Mission Directorate

EVA—extravehicular activity

Exosphere—the highest layer of an atmosphere

feldspathic—pertaining to rocks rich in feldspar minerals

Ga—one billion years

geochronology—determination of the time at which a rock crystallized, usually by radioactive decay of parent-daughter isotope pairs: U-Pb, Sm-Nd, K-Ar, or Rb-Sr

hyperthermophile—microorganisms that live in hot environments, above 60°C

igneous rocks—rocks crystallized from a magma

ISRU—in situ resource utilization

Isua greenstone rocks—a geological formation in southwestern Greenland (Isua) composed of ancient surface rocks

KREEP—lunar basalts and breccias that are rich in potassium (K), rare-earth elements (REE), and phosphorus (P)

Kuiper Belt—a region of the solar system distributed in a roughly circular disk extending from 40 to 100 astronomical units from the Sun

Late Heavy Bombardment (LHB) hypothesis—the theory that meteorite bombardment of the inner solar system declined after accretion of the planets and then peaked again at 3.9 Ga. This event is proposed to have limited the emergence of life on Earth

LCROSS—the Lunar Crater Observation and Sensing Satellite, a secondary payload to be launched with the Lunar Reconnaissance Orbiter

LEAG—the NASA Lunar Exploration Analysis Group (LEAG) is responsible for analyzing scientific, technical, commercial, and operational issues associated with lunar exploration

LPRP—the Lunar Precursor and Robotic Program of the Exploration Systems Mission Directorate

LRO—NASA's Lunar Reconnaissance Orbiter

LSAM—NASA's Lunar Surface Access Module

Lunar A—a Japanese Aerospace Exploration Agency lunar mission

lunar dynamo—for the ancient Moon, the possibility that a lunar magnetic field might have been generated by the internal motion of an iron-rich molten core

Lunar Magma Ocean hypothesis—the hypothesis that when the Moon formed, it was molten to a depth of hundreds of kilometers; its crystallization produced the primary crust and mantle

mafic materials—a dark-colored, igneous rocks rich in ferromagnesian minerals

magma—molten rock from the interior of a planetary body

mantle—the geologic zone above the core and below the crust

MER—NASA's Mars Exploration Rover

obliquity—the angle between the orbital plane of an object and the plane of its rotational equator

olivine—a magnesium iron silicate mineral $(FeMg)_2SiO_4$

parautochthonous—a rock type intermediate in tectonic character between those found at the site of their formation and those that come from another site

Permian/Triassic boundary—on Earth, the layer of rocks between the Permian and Triassic periods, about 251 million years ago, the time of a major extinction event

petrology—a branch of geology dealing with the composition, mineralogy, origin, occurrence, history, and structure of rocks

plagioclase—a type of feldspar and one of the most common rock-forming minerals, $(CaNa)(AlSi)_4O_8$

pluton—a large body of igneous rock created by the subsurface intrusion of magma

pre-Nectarian period—the lunar geologic period from the formation of the Moon to about 3.9 Ga ago

protolith—the original rocks from which igneous, metamorphic, or sedimentary rocks were formed

pyroclastic—a type of rock material formed by volcanic explosion

pyroxene—a group of common ferromagnesian rock-forming minerals

regolith—on the Moon, the surface rock debris that overlies bedrock

remanent magnetization—also called paleomagnetism, the component of a rock's magnetism that has a fixed direction and is independent of Earth's magnetic field

rheology—the study of the deformation and flow of matter

SBE—surface boundary exosphere

SELENE—Selenological and Engineering Explorer, a Japanese Aerospace Exploration Agency lunar orbiter

SIM—the Apollo Scientific Instrument Module, which contained panoramic and mapping cameras, a gamma-ray spectrometer, a laser altimeter, and a mass spectrometer

SIMS—secondary-ion mass spectrometry

SMART-1—Small Missions for Advanced Research in Technology, a European Space Agency lunar mission

SMD—NASA's Science Mission Directorate

SPA—the lunar South Pole-Aitken Basin

stratigraphy—the study of rock layers

TEM—transmission electron microscopy

U-Pb geochronology—the determination of the age of a rock based on radioactive decay of isotopes of U and Th to Pb, usually in the mineral zircon

VSE—NASA's Vision for Space Exploration

xenolith—a foreign inclusion in an igneous rock

zircon—a silicate mineral ($ZrSiO_4$)

C

Public Agendas for Meetings

June 20, 2006

Closed Session

8:00 a.m.   Discussion

Open Session

11:00 a.m.   Welcome Guests, Introductions

11:05 a.m.   Talk 1—Perspective from Science Mission Directorate (SMD) Program Management
                Paul Hertz, NASA Science Mission Directorate Chief Scientist

Noon   Lunch

1:00 p.m.   Talk 2—Lunar Sample Science and Sample Return
                Charles (Chip) Shearer, University of New Mexico

2:00 p.m.   Talk 3—Lunar Sample Curation
                Gary Lofgren, NASA Johnson Space Center

3:00 p.m.   Break

3:15 p.m.   Talk 4—Lunar Exploration Analysis Group (LEAG)
                G. Jeffrey Taylor, University of Hawaii (teleconference)

4:15 p.m.   Discussion of Presentations

5:00 p.m.   Adjourn

5:45 p.m.   Dinner for Committee and Speakers

June 21, 2006

Open Session

8:30 a.m.   Welcome

8:35 a.m.   Talk 5—Lunar Geophysical Network
             *Clive Neal, Associate Professor of Civil Engineering and Geological Sciences,
             University of Notre Dame (teleconference)*

9:35 a.m.   Talk 6—Lunar Geophysics
             Norman Sleep, Stanford University

10:35 a.m.   Break

10:55 a.m.   Talk 7—South Pole-Aitken Basin
             Brad Jolliff, Washington University at St. Louis

Noon   Lunch

1:00 p.m.   Discussion of Presentations

2:00 p.m.   Discussion of White Papers

3:00 p.m.   Break

3:15 p.m.   Discussion of Critical Near-Term Lunar Science Issues
             Discussion of the Outline of the Interim Report

4:00 p.m.   Overnight Assignments

4:15 p.m.   Talk 8—History of Lunar Science
             S. Ross Taylor, Australian National University (teleconference)

5:15 p.m.   Adjourn

June 22, 2006

Closed Session

8:00 a.m.   Discussion

Noon   Adjourn

MEETING 2 AGENDA

August 2, 2006

Closed Session

8:00 a.m.   Discussion

Open Session

11:45 a.m.   Lunch

12:45 p.m.   Possible Lunar Science Mission Architectures
               Butler Hine, NASA Ames Research Center

1:30 p.m.   Lunar Science Objectives of Lander Missions
               Paul Spudis, Johns Hopkins University

2:15 p.m.   Science and NASA's Vision: Lessons Learned from Apollo
               James Head III, Brown University (teleconference)

3:00 p.m.   Break

3:15 p.m.   Lunar Robotics and Telerobotics
               Ayanna Howard, Georgia Institute of Technology

4:00 p.m.   Science and NASA's Vision: Lessons Learned from Apollo
               Noel Hinners, Lockheed-Martin (retired)

4:45 p.m.   Discussion of Presentations

5:15 p.m.   Adjourn

6:00 p.m.   Dinner

August 3, 2006

Closed Session

8:00 a.m.   Discussion

5:00 p.m.   Adjourn

August 4, 2006

Closed Session

8:00 a.m.   Discussion

Noon   Adjourn

MEETING 3 AGENDA

October 25, 2006

Closed Session

8:00 a.m.   Discussion

Open Session

9:00 a.m.   Update from NASA/Exploration Systems Mission Directorate (ESMD)
              Michael Wargo, NASA ESMD

9:45 a.m.   Break

10:00 a.m.   Thoughts on the Vision and Lunar Science Opportunities
              Simon P. Worden, NASA Ames Research Center

10:45 a.m.   Feedback on the Interim Report from NASA/SMD
              Robert Fogel, NASA Science Mission Directorate (teleconference)

11:30 a.m.   Advanced Robotics and Telerobotics for Lunar Exploration
              David Lavery, NASA SMD, and Rob Ambrose, NASA Johnson Space Center

12:15 p.m.   General Discussion

12:30 p.m.   Lunch

1:30 p.m.   Dusty Plasmas on the Moon
              Timothy Stubbs, University of Maryland

2:15 p.m.   Lunar Volatiles
              Richard R. Vondrak, NASA Goddard Space Flight Center

3:00 p.m.   Break

3:15 p.m.   Science Opportunities from Lunar Reconnaissance Orbiter
              Gordon Chin, NASA Goddard Space Flight Center

4:00 p.m.   General Discussion

Closed Session

7:00 p.m.   Dinner

October 26, 2006

Closed Session

8:00 a.m.   Discussion

Open Session

9:45 a.m.	The Changing Environment at the Moon *Joseph Borovsky, Los Alamos National Laboratory*
10:30 a.m.	Astronomy from the Moon *Daniel Lester, University of Texas*
11:15 a.m.	Magnetospheric Measurements from the Moon *James Burch, Southwest Research Institute*
Noon	Lunch
1:00 p.m.	Earth Remote Sensing from the Moon *Francisco Valero, Scripps Institute of Oceanography*
1:45 p.m.	The Archean and Hadean Earth *John Valley, University of Wisconsin*
2:30 p.m.	General Discussion of Presentations
3:00 p.m.	Break

Closed Session

3:15 p.m.	Committee's Final Report
4:30 p.m.	Discussion of Priorities
6:00 p.m.	Adjourn

October 27, 2006

Closed Session

8:00 a.m.	Discussion
Noon	Adjourn

MEETING 4 AGENDA

February 13, 2007

Closed Session

8:00 a.m.	Discussion
5:00 p.m.	Committee Dinner

February 14, 2007

Open Session

8:30 a.m.   Workshop on Astrophysics from the Moon
                   Mario Livio, Space Telescope Science Institute

9:15 a.m.   Discussion with Workshop Attendees

9:45 a.m.   Break

Closed Session

10:00 a.m.   Discussion

Open Session

1:00 p.m.   Preview of the NASA Advisory Council's Lunar Science Workshop
                   Brad Jolliff, Washington University at St. Louis

2:00 p.m.   General Discussion

2:30 p.m.   Break

Closed Session

2:45 p.m.   Discussion

5:00 p.m.   Adjourn

February 15, 2007

Open Session

8:00 a.m.   Radio Astronomy on the Moon
                   Jack O. Burns, University of Colorado, Boulder

9:00 a.m.   General Discussion

9:30 a.m.   Break

Closed Session

9:45 a.m.   Discussion

Noon   Adjourn

D

Lunar Beijing Declaration

The following declaration is reprinted, courtesy of the International Lunar Exploration Working Group (ILEWG), http://sci.esa.int/ilewg.

8TH ILEWG (INTERNATIONAL LUNAR EXPLORATION WORKING GROUP) INTERNATIONAL CONFERENCE ON EXPLORATION AND UTILIZATION OF THE MOON

Lunar Beijing Declaration

More than 240 experts and 300 students from 18 countries met in Beijing from 23 to 27 July 2006 for the 8th ILEWG Conference on Exploration and Utilization of the Moon, kindly and effectively hosted by CNSA [China National Space Administration], with support from CASC, LEPC, CSSAR, and CAECC. Based on the deliberations and opinions, the participants have prepared the Lunar Beijing Declaration.

We salute the SMART-1 team for a successful technology and science mission, as the spacecraft approaches its grand finale. This small spacecraft has initiated an exciting International Lunar Decade that will inspire a new generation of lunar explorers.

Within the next two years, four independent spacecraft (SELENE, Chang'e 1, Chandrayaan 1 and Lunar Reconnaissance Orbiter) will orbit the Moon carrying an extensive array of sophisticated science and exploration instruments. Our understanding of the Moon and its resources will be revolutionized when the rich array of data from this flotilla is analyzed by scientists and experts around the world.

Since the first phase of lunar exploration is centered on remote-sensing observations, we endorse the following actions as being of long-term mutual benefit:

1. Internationally coordinated analyses should be carried out to facilitate the validation of data sets produced by different instruments and to enhance the usefulness of information acquired by multiple spacecraft

2. A small number of specific targets are recommended to facilitate both the cross-calibration of different instruments and to train young explorers in lunar science issues. After initial calibration, data should be made available for coordinated analyses by the international community

3. All solar monitor data from lunar orbital missions should, to the extent possible, be made available as rapidly as possible. Cross correlation of this information will improve calibration of all the instruments dependent on knowledge of solar fluxes

4. Every effort should be made to coordinate development and utilization of a common, improved Lunar

Coordinates Reference Frame

5. Lunar mission teams should archive final mission data products in a PDS-compatible form, to implement international standards for access, and to support Unicode, or other necessary format

6. The establishment of Common standards for S-band spacecraft communication, with potential for common tracking operations and backup support to other missions, if necessary

7. A coordinated campaign to provide data cross-check and validation for modern-era missions that have overlap in coverage, with data and experience from Past missions (including archived and digitized Apollo and Soviet-era lunar data) is recommended

8. Information about the five impact events/probes (SMART-1, Chandrayaan-1, LCROSS, SELENE RSAT, and VSAT) and subsequent impacts of lunar crafts should be coordinated with other space missions. Ground- and space-based measurements should be conducted for near-side events. All of the planned four orbital missions are asked observe the SMART-1 impact site. Before, after, and real-time measurements should be planned by all spacecraft that are in orbit during the impact events

To strengthen exchange between lunar experts and to enhance collaboration, we recommend to international science and space organizations join in and support the International Lunar Decade.

For the subsequent phase of Lunar Global Robotic Village and preparation for human exploration, we further recommend:

9. To promote use of standardized telecommunications, navigation, and VLBI support for future orbiter, lander and rover missions. We propose that ILEWG and agencies study the opportunity to embark some payload technologies for navigation and guidance on orbiters and landers as part of a Global Moon Navigation and Positioning System

10. Lunar Missions should document their plans for end of operations. Before completing their mission, future orbiters could be placed on frozen stable orbits where they can participate in a joint infrastructure for data relay, aid to navigation and lunar internet, in addition to landed surface beacons

11. Recognizing the importance of the geophysical studies of the interior of the Moon for understanding its formation and evolution, the necessity for a better monitoring of all natural hazards (radiation, meteorites impacts and shallow moonquakes) on the surface, and the series of landers planned by agencies in the period 2010-2015 as an unique opportunity for setting up a geophysical network on the Moon, we recommend the creation of an international scientific working group for definition of a common standard for future Moon network instruments, in a way comparable to Earth seismology and magnetism networks. Interested agencies and research organizations should study inclusion of network instruments in the Moon landers payload and also piggyback deployment of a Moon Geophysical and Environmental Suitcase

12. The importance of protecting the Moon becomes more urgent than ever before, as we enter a decade with many planned lunar exploration missions, including orbiters, impactors, penetrators and landers. Space agencies should give their attention to the protection of the Moon for sustainable exploration, research and utilization. A dedicated task force should be set up to study this issue and produce a recommendation for all future missions

13. Lunar Exploration is ideal for outreach activities that are accessible and inspiring for the next generation of explorers. Students should work on lunar payloads and participate in missions. We propose to use milestones of lunar missions for public outreach events promoting exploration, space science and technology

We reaffirm our commitment, with the international lunar missions and research community, to prepare the way for global participation in the extension of human presence on the Moon and beyond, for the benefit of all mankind.

Beijing, July 27, 2006
Unanimously approved by the participants

E

Committee Outreach Activities

TABLE E.1 Summary of Outreach Presentations of the Committee on the Scientific Context for Exploration of the Moon

Date	Event	Location	Presentation Type	Presenters
2006				
July 24-27	8th ILEWG International Conference on Exploration and Utilization of the Moon	Beijing, China	Oral presentation	David H. Smith
September 18-22	European Planetary Science Congress (Europlanet)	Berlin, Germany	Oral presentation	Harald Hiesinger
September 25	NASA Advisory Council Planetary Science Subcommittee	Boulder, Colorado	Oral presentation	George Paulikas, Carlé Pieters
October 3-4	Lunar Reconnaissance Orbiter Camera Team Meeting	Phoenix, Arizona	Conference call	Harald Hiesinger
October 8-13	38th Annual Division for Planetary Sciences Meeting	Pasadena, California	Poster paper	David H. Smith, George Paulikas
			Press conference; forum	George Paulikas, Carlé Pieters, Bruce Banerdt
October 11	NRC Committee on Solar and Space Physics	Washington, D.C.	Conference call	George Paulikas, Carlé Pieters
October 31-November 2	Space Resources Roundtable VIII	Golden, Colorado	Oral presentation	Michael Duke
November 29	NRC Committee on Astronomy and Astrophysics	Irvine, California	Oral presentation	George Paulikas, Carlé Pieters

TABLE E.1 *continued*

Date	Event	Location	Presentation Type	Presenters
November 28-30	Space Telescope Science Institute (STScI) Astrophysics Enabled by Return to the Moon Workshop	Baltimore, Maryland	Oral presentation	Daniel Lester
December 4-6	American Institute of Aeronautics and Astronautics 2nd Space Exploration Workshop	Houston, Texas	Panel discussion	Noel Hinners
December 4	NRC Committee on Planetary and Lunar Exploration	Irvine, California	Oral presentation	George Paulikas, Carlé Pieters
December 11-15	American Geophysical Union	San Francisco, California	Panel discussion, oral presentation and display	Carlé Pieters
2007 January 5-10	209th American Astronomical Society Meeting	Seattle, Washington	Display	n/a
February 19	NRC Committee on the Origins and Evolution of Life	Washington, D.C.	Conference call	George Paulikas, Carlé Pieters
February 21-23	International Space University 11th Annual International Symposium	Strasbourg, France	Oral presentation	Harald Hiesinger
February 26-March 2	NASA Advisory Council Lunar Science Workshop	Tempe, Arizona	Oral presentation	Carlé Pieters
March 12-16	38th Lunar and Planetary Science Conference	Houston, Texas	Forum	Carlé Pieters
April 15-20	European Geosciences Union General Assembly	Vienna, Austria	Oral presentation	Harald Hiesinger

F

Biographies of Committee Members and Staff

GEORGE A. PAULIKAS, *Chair*, has been at the forefront of advances in space science and space systems and has made many technical contributions to the development of national security space systems. He retired after 37 years at the Aerospace Corporation, having joined Aerospace in 1961 as a member of the technical staff and later becoming department head, laboratory director, vice president, and senior vice president. He became executive vice president in 1992. He received the company's highest award, the Trustees' Distinguished Achievement Award, in 1981 in recognition of research leading to a new understanding of the dynamics of space radiation and its effect on spacecraft. Dr. Paulikas was vice chair of the National Research Council's (NRC's) Space Studies Board (SSB) from 2003 to 2006. He has also served on a number of NRC study committees, including the Committee on an Assessment of Balance in NASA's Science Programs (vice chair), the Committee on the Scientific Context for Space Exploration, the Committee on Systems Integration for Project Constellation, the Workshop Committee on Issues and Opportunities Regarding the Future of the U.S. Space Program, and the Committee to Review the NASA Earth Science Enterprise Strategic Plan.

CARLÉ M. PIETERS, *Vice Chair*, is a professor in the Department of Geological Sciences at Brown University. Her research areas include the study of lunar evolution, asteroid-meteorite links, space weathering of materials, Mars mineralogy, and planetary exploration; she also conducts laboratory spectroscopy experiments on planetary materials. She was a member of the science team for the DOD-NASA Clementine mission. Dr. Pieters is active in international cooperation on lunar research with European, Russian, and Indian scientists. She is principal investigator (PI) of a Discovery Mission of Opportunity for the Moon Mineralogy Mapper to be flown on Chandrayaan-1, India's mission to the Moon to be launched in 2008. Her NRC service includes membership on the SSB, the Committee on Planetary and Lunar Exploration, the Study Team on the Terrestrial Planets, and the Task Group on Research Prioritization, and as chair of the Inner Planets Panel of the Committee on a New Science Strategy for Solar System Exploration.

WILLIAM B. BANERDT is a principal research scientist at the California Institute of Technology's Jet Propulsion Laboratory. His research interests cover planetary science, geophysics, gravity, and seismology. He serves as project scientist for the Mars Exploration Rovers project, is a participating scientist for the Mars Orbiter Laser Altimeter Investigation on the Mars Global Surveyor, and is a co-investigator on the Rosetta Surface Electrical, Seismic and Acoustic Monitoring Experiment Team. Dr. Banerdt served as PI on the Discovery Mars NetLander Project (2001-2003), PI on the NetLander Short-Period Seismometer (1998-2003), guest investigator on the Magellan Radar Investigation Team (1990-1994), and chair of the NASA Mars Data Analysis Program Review Panel (2004-2005). He currently serves on the NRC Committee on Planetary and Lunar Exploration.

JAMES L. BURCH is a vice president at the Southwest Research Institute (SwRI) in the Space Science and Engineering Division. Dr. Burch was a space physicist at NASA for 6 years prior to his taking a position at SwRI in 1977. In 1996, he was selected as the PI for the NASA Imager for Magnetopause-to-Aurora Global Exploration mission, which provided global images of key regions of Earth's magnetosphere as they respond to variations in the solar wind. He now serves as PI and chair of the Science Working Group for the NASA Magnetospheric Multiprobe mission. Dr. Burch was elected a fellow of the American Geophysical Union in recognition of his work in the field of space physics and aeronomy, including research on the interaction of the solar wind with Earth's magnetosphere and the physics of the aurora. He recently served on the governing board of the American Institute of Physics and previously chaired the NRC Committee on Solar and Space Physics. He also served on the NRC Committee for the Review of NASA Science Mission Directorate Science Plan.

ANDREW CHAIKIN is a science journalist, a space historian, and a commentator for National Public Radio, and has been an adviser to NASA on space policy and public communications. Mr. Chaikin has authored books and articles about space exploration and astronomy for more than two decades. He is also active as a lecturer at museums, schools, corporate events, and in radio and television appearances. He is best known as the author of *A Man on the Moon: The Triumphant Story of the Apollo Space Program*, first published in 1994. Mr. Chaikin spent 8 years writing and researching *A Man on the Moon*, including hundreds of hours of personal interviews with the 23 surviving lunar astronauts. He co-edited *The New Solar System*, a compendium of writings by planetary scientists, and he is also the author of *Air and Space: The National Air and Space Museum Story of Flight*. He collaborated with moonwalker-turned-artist Alan Bean to write *Apollo: An Eyewitness Account,* and he co-authored the text for the collection of Apollo photography, *Full Moon*. Mr. Chaikin served on the Viking missions to Mars team at NASA's Jet Propulsion Laboratory, and he was a researcher at the Smithsonian's Center for Earth and Planetary Studies.

BARBARA A. COHEN is a research assistant professor and assistant curator of meteorites in the Institute of Meteoritics at the University of New Mexico. Her research efforts combine geochemistry and geochronology of terrestrial, lunar, and meteoritic samples to contribute to the understanding of planetary surface processes, including impact processing, igneous magmatism, and aqueous alteration. She has served on the Curation and Analysis Planning Team for Extraterrestrial Materials (CAPTEM) and the CAPTEM lunar sample subcommittee. She is currently a member of the Athena Science team for the Mars Exploration Rovers mission.

MICHAEL DUKE is a planetary geologist who recently retired as the director of the Center for Commercial Applications of Combustion in Space at the Colorado School of Mines. His principal research focuses on the general area of study that relates to the use of in situ resources to support human exploration missions to the Moon and Mars. His planetary science interests relate to the mineralogy and petrology of meteorites and lunar materials. Dr. Duke worked at the NASA Johnson Space Center for 25 years prior to accepting the position at the Colorado School of Mines in 1998. He has also been a research scientist at the U.S. Geological Survey (1963-1970) and curator of NASA's lunar sample collection (1970-1977). Dr. Duke received the NASA Exceptional Scientific Achievement Award (1973) and the AIAA's Space Science Medal (1998), and he was a Distinguished Federal Executive (1988). He served as a member of the NRC Panel on Solar System Exploration of the Committee on Priorities for Space Science Enabled by Nuclear Power and Propulsion.

ANTHONY W. ENGLAND, University of Michigan, resigned from the committee on August 11, 2006, because of other commitments.

HARALD HIESINGER is a professor of geological planetology at Westfälische Wilhelms-Universität, Münster. He was formerly a senior research associate in geological sciences at Brown University and an assistant professor of planetary sciences at Central Connecticut State University. His research focuses on the new imaging data of the martian surface provided by the Mars Express High Resolution Stereo Camera and on dating lunar mare basalts to investigate the thermal and mineralogical evolution of the Moon and their consequences for volcanism. He is co-investigator on the Lunar Reconnaissance Orbiter, scheduled for launch in 2008. Dr. Hiesinger is currently involved in three international space missions—the Lunar Reconnaissance Orbiter, the European Mars Express, and the European Bepi Colombo. He has published many scientific papers on the Moon, Mars, and Ganymede and has written and/or contributed to two book chapters.

NOEL W. HINNERS is a senior research associate at the Laboratory for Atmospheric and Space Physics and lecturer on space policy at the University of Colorado. He retired in January 2002 from Lockheed Martin Astronautics where he was vice president of light systems with responsibility for NASA's Mars Global Surveyor, the Mars Reconnaissance Orbiter, and the Stardust and Genesis Discovery missions. Dr. Hinners served as an associate deputy administrator and the NASA chief scientist (1987-1989), director of the NASA Goddard Space Flight Center (1982-1987), director of the Smithsonian's National Air and Space Museum (1979-1982), NASA associate administrator for space science (1974-1979), and director of NASA's Lunar Programs (1972-1974). He served on the NRC Committee on Technology for Human/Robotic Exploration and Development of Space (2001), the Committee on Human Exploration (chair, 1996-1997), and the Committee on Human Exploration (chair, 1990-1993).

AYANNA M. HOWARD is an associate professor of systems and controls in the School of Electrical and Computer Engineering at the Georgia Institute of Technology (Georgia Tech) and is the founder of the Human-Automation Systems Laboratory at Georgia Tech. From 1993 to 2005, Dr. Howard was at NASA's Jet Propulsion Laboratory (JPL) where she led research efforts on various robotic projects utilizing vision, fuzzy logic, and neural network methodologies. As a robotic scientist at JPL, she specialized in the study of artificial intelligence. Her area of research at Georgia Tech focuses on the concept of humanized intelligence—the process of embedding human cognitive capability into the control path of autonomous systems. This work, which addresses issues of autonomous control as well as aspects of interaction with humans and the surrounding environment, has resulted in more than 60 publications covering topics ranging from autonomous rover navigation for planetary surface exploration to intelligent terrain assessment algorithms for landing on Mars. Dr. Howard's unique accomplishments have been documented in more than 12 featured articles, and she was named as one of the world's top young innovators of 2003 by Massachusetts Institute of Technology's *Technology Review* and in *TIME*'s "Rise of the Machines" article in 2004.

DAVID J. LAWRENCE is a technical staff member in the Space Science and Applications Group at the Los Alamos National Laboratory (LANL) and recently served as acting director of the LANL Center for Space Science and Exploration, which oversees all NASA-funded programs at LANL. Dr. Lawrence specializes in the study of planetary compositions using various nuclear detection techniques as well as in Department of Energy (DOE)-sponsored treaty verification efforts. He has extensive instrumentation and data analysis experience for spaceflight missions funded by NASA and DOE. His involvement in NASA missions includes Deep Space-1, Lunar Prospector, Mars Odyssey, and MESSENGER, as well as the European Space Agency's SMART-1 mission. Dr. Lawrence is PI for two neutron instruments that will fly on DOE-funded space-based treaty-monitoring missions. He is PI or co-investigator on numerous NASA-funded planetary science and instrument development grants and has also served on various NASA review panels. Dr. Lawrence is the author or co-author of more than 60 peer-reviewed publications in the areas of space and planetary sciences and instrumentation. He served on the NRC Committee for the Review of NASA's Capability Roadmaps Panel A: In-Situ Resource Utilization (2005).

DANIEL F. LESTER is a research scientist at the McDonald Observatory of the University of Texas. His research specialty is infrared studies of star formation in galaxies. Dr. Lester was previously a staff scientist at the University of Hawaii's Institute for Astronomy. He has also worked closely on the conceptual development of the Stratospheric Observatory for Infrared Astronomy and has been active in community strategic planning and policy development for space astronomy. Dr. Lester was the PI and team leader for NASA's Single Aperture Far Infrared

"Vision Mission" study. He served on the NRC Panel on Astronomy and Astrophysics of the Committee on Priorities for Space Science Enabled by Nuclear Power and Propulsion. He currently serves on the Astronomy and Astrophysics Advisory Committee, a multiagency oversight committee for astronomy.

PAUL G. LUCEY is a professor in the Department of Geology and Geophysics and the Hawaii Institute of Geophysics and Planetology at the University of Hawaii. His research focuses on the development of remote sensing instruments. His planetary science interests focus on the study of the Moon and asteroids, but he has also conducted research and published papers on Mercury, Venus, and Mars. Dr. Lucey's lunar research has primarily involved the examination of remote sensing data as it relates to the composition of the lunar crust and the surfaces of asteroids. His spacecraft experience includes being a member of the science team on the Department of Defense/NASA Clementine lunar orbiter and a participating scientist for NASA's Near-Earth Asteroid Rendezvous mission. Dr. Lucey also served as one of the lead authors for a white paper drafted by the NASA Astrobiology Institute entitled "Astrobiology Science Goals and Lunar Exploration."

S. ALAN STERN, Southwest Research Institute, resigned from the committee on September 24, 2006, to join the NASA Advisory Committee Science Subcommittee (and on April 2, 2007, became Associate Administrator for NASA's Science Mission Directorate).

STEFANIE TOMPKINS is a deputy operations manager at Science Applications International Corporation, where she manages research and development projects for hyperspectral image exploitation and conducts independent research as a NASA PI. Her research efforts focus on reflectance spectroscopy as it applies to remote geochemical analysis of planetary surfaces, including Earth's Moon and Earth itself. Her most recent publications discuss her work in mapping the subsurface composition of the Moon using spectroscopic signatures of rocks excavated by impact craters and in algorithm development for modeling mixed pixels in spectral imagery.

FRANCISCO P.J. VALERO is the associate director of the Center for Atmospheric Sciences at the Scripps Institution of Oceanography (SIO) at the University of California, San Diego. His research interests include climate studies, atmospheric radiative transfer, and solar system exploration. Dr. Valero joined SIO in 1993 after many years as a senior scientist at the NASA Ames Research Center, where he began his career as a specialist in spectroscopy of stellar and planetary atmospheres. While at Ames, he pioneered advanced radiometric instrumentation for climate studies from research aircraft. Dr. Valero's instruments continue to play a role in many meteorological and climate experiments led by NASA and the Department of Energy. He was the PI for the TIREX experiment in NASA's Comet Rendezvous Asteroid Flyby and NASA's Deep Space Climate Observatory satellite mission that was built to make high-time-resolution multispectral observations of the entire Earth disk from the Lagrange L-1 point.

JOHN W. VALLEY is the Charles R. Van Hise Professor of Geology in the Department of Geology and Geophysics at the University of Wisconsin-Madison. His primary areas of interest are in stable isotope geochemistry, with particular emphasis on the evolution of Earth's crust. In 2005, he established WiscSIMS, a laboratory dedicated to in situ microanalysis of stable isotope ratios with applications to Earth, materials, and biological sciences. In 2003, Dr. Valley received the American Geophysical Union Bowen Award, presented by the Volcanology, Geochemistry, and Petrology Section for work on zircons from early Archean rocks of northwestern Australia, which provide documentation of previously missing Earth history with evidence of an early ocean and a relatively cool history during the Hadean Era. In 2005 to 2006, he was president of the Mineralogical Society of America.

CHARLES D. WALKER is an independent consultant and the former director of NASA Systems Government Relations at the Boeing Company. From 1979 to 1986, he was chief test engineer and payload specialist for the McDonnell Douglas Electrophoresis Operations in Space (EOS) commercialization project, where he led the EOS laboratory test and operations team in developing biomedical products. His contributions to the program included engineering planning, design and development, product research, and spaceflight and the evaluation of the continuous flow electrophoresis system (CFES) device. He was responsible for training NASA astronaut crews in the operation of the CFES payload on STS-4, STS-6, STS-7, and STS-8. Although never an employee of NASA, Mr. Walker was confirmed in 1983 as the first industry payload specialist, and he accompanied the McDonnell Douglas CFES equipment as a crew member on Space Shuttle missions 41-D, 51-D, and 61-B. In May 1986, Mr. Walker was appointed as a special assistant to the president of McDonnell Douglas Space Systems Company. He served on the NRC Steering Committee for Workshops on Issues of Technology Development for Human and Robotic Exploration and Development of Space and on the Space Applications Board.

NEVILLE J. WOOLF is a professor in the Department of Astronomy at the University of Arizona and director of the university's Life and Planets Astrobiology Center. His research interests focus on astronomical instrumentation. He is particularly interested in the development of interferometric techniques and in their application in the search for extrasolar planets. His recent research has included studies of the potential use of the lunar surface as an astronomical observing location. He has a longstanding interest in the characterization of the spectra of planetary bodies and in the identification of spectral features in planetary atmospheres that may be indicative of the presence of life. He is currently a member of the NRC Committee on the Origins and Evolution of Life, and he served as a member of the Committee on Planetary Protection Requirements for Venus Missions (2005) and the Committee on the Astrophysical Context of Life (2003-2004).

Staff

ROBERT L. RIEMER, *Study Director*, joined the staff of the NRC in 1985. He is a senior program officer and served in that capacity for the two most recent decadal surveys of astronomy and astrophysics and has worked on studies in many areas of physics and astronomy for the Board on Physics and Astronomy (where he served as associate director from 1988 to 2000) and the SSB. Prior to joining the NRC, Dr. Riemer was a senior project geophysicist with Chevron Corporation. He received his Ph.D. in experimental high-energy physics from the University of Kansas-Lawrence and his B.S. in physics and astrophysics from the University of Wisconsin-Madison.

DAVID H. SMITH joined the staff of the SSB in 1991. He is the senior staff officer and study director for a variety of NRC activities, including the Committee on Planetary and Lunar Exploration, the Committee on the Origins and Evolution of Life, the Mars Astrobiology Task Group, the Mars Architecture Assessment Task Group, the Committee on the Limits of Organic Life in Planetary Systems, the Task Group on Organic Environments in the Solar System, the Nuclear Systems Committee, and the proposed Lunar Science Strategy Committee. He also organizes the SSB's summer intern program and supervises most, if not all, of the interns. He received a B.Sc. in mathematical physics from the University of Liverpool in 1976 and a D.Phil. in theoretical astrophysics from Sussex University in 1981. Following a postdoctoral fellowship at Queen Mary College, University (1980-1982), he held the position of associate editor and, later, technical editor of *Sky and Telescope*. Immediately prior to joining the staff of the SSB, Dr. Smith was a Knight Science Journalism Fellow at the Massachusetts Institute of Technology (1990-1991).

STEPHANIE BEDNAREK, working as a research assistant during her SSB space policy internship, is attending University of Virginia (UVA) for a bachelor of science degree in aerospace engineering with a minor in astronomy. Ms. Bednarek has worked as an intern with Aerospace Industries Association and Orbital Sciences Corporation. At UVA, she served as the student director of engineering visitation and undergraduate recruitment and secretary of UVA's American Institute of Aeronautics and Astronautics student chapter. In addition, she served as vice president of UVA's Equestrian Team and is a member of Alpha Delta Pi sorority. After graduation in May 2007, she plans to attend graduate school to study science and technology and policy and to pursue a career in space policy.

CATHERINE A. GRUBER, an assistant editor for the SSB, joined the board as a senior program assistant in 1995. Ms. Gruber first came to the NRC in 1988 as a senior secretary for the Computer Science and Telecommunications Board and has also worked as an outreach assistant for the National Academy of Sciences-Smithsonian Institution's National Science Resources Center. She was a research assistant (chemist) in the National Institute of Mental Health's Laboratory of Cell Biology for 2 years. She has a B.A. in natural science from St. Mary's College of Maryland.

RODNEY N. HOWARD joined SSB as a senior project assistant in 2002. Before joining SSB, most of his vocational life was spent in the health profession as a pharmacy technologist at Doctor's Hospital in Lanham, Maryland, and as an interim center administrator at the Concentra Medical Center in Jessup, Maryland. During that time, he participated in a number of Quality Circle Initiatives that were designed to improve relations between management and staff. Mr. Howard obtained his B.A. in communications from the University of Baltimore County in 1983.